Computer System Reliability

Safety and Usability

T0179165

Computer System Reliability

Safety and Usability

B. S. DHILLON

CRC Press
Taylor & Francis Group
Boca Raton London New York

CRC Press is an imprint of the
Taylor & Francis Group, an **informa** business

CRC Press
Taylor & Francis Group
6000 Broken Sound Parkway NW, Suite 300
Boca Raton, FL 33487-2742

First issued in paperback 2019

© 2013 by Taylor & Francis Group, LLC
CRC Press is an imprint of Taylor & Francis Group, an Informa business

No claim to original U.S. Government works

ISBN-13: 978-0-367-38006-9

Library of Congress Cataloging-in-Publication Data

Dhillon, B. S.
    Computer system reliability : safety and usability / author, B.S. Dhillon.
    pages cm
    Includes bibliographical references and index.

    1. Computer systems--Reliability. 2. Fault-tolerant computing. 3. Electronic digital computers--Reliability. I. Title.

    QA76.76.R44D495 2013
    004.2--dc23   2013008020

**Visit the Taylor & Francis Web site at
http://www.taylorandfrancis.com**

**and the CRC Press Web site at
http://www.crcpress.com**

This book is affectionately dedicated to all my Scythian ancestors

for their firm belief in number 40. (This is my 40th book!)

Contents

Preface

Computer systems are increasingly being used at an alarming rate for various purposes. They have become an important element of the world economy because billions of dollars are spent each year to develop, manufacture, operate, and maintain various types of computer systems around the globe. Their reliability, safety, and usability have become an important concern because of problems such as high cost, wrong decisions and actions, and accidental deaths. For example, a study performed by the National Institute of Standards in 2002 found that software defects alone cost the United States economy about $59 billion annually, i.e., around 0.6% of its gross domestic product (GDP).

Computer system reliability, safety, and usability have become more important than ever before. In response, a large number of journal and conference proceedings' articles on various aspects of computer system reliability, safety, and usability have been published over the years. However, to the best of the author's knowledge, there is no specific book on the topic. This causes a great deal of difficulty for information seekers because they have to consult many different and diverse sources.

Thus, the main objective of this book is to combine computer system reliability, safety, usability, and other related topics into a single volume and to eliminate the need to consult many different and diverse sources to obtain desired information. The book contains a chapter on mathematical concepts considered necessary to understand the material presented in subsequent chapters.

The topics covered in the volume are treated in such a manner that the reader will require no previous specialized knowledge to understand the contents. At appropriate places, the book contains examples along with their solutions, and at the end of each chapter there are numerous problems to test the reader's comprehension. The sources of most of the materials presented are given in the reference section at the end of each chapter. An extensive list of publications dating from 1967 to 2011—directly or indirectly related to computer system reliability, safety, and usability—is provided at the end of this book to give readers a view of the intensity of the developments in this area.

The book is composed of 11 chapters. Chapter 1 presents various introductory aspects of computer system reliability including safety; usability-related facts, figures, terms, and definitions; and sources for obtaining useful information on computer system reliability, safety, and usability. Chapter 2 reviews mathematical concepts considered useful to understanding subsequent chapters. Some of the topics covered in the chapter are arithmetic mean and mean deviation, Boolean algebra laws, probability properties, probability distributions, and useful definitions.

Chapter 3 presents various introductory aspects of reliability, safety, and usability. Chapter 4 presents computer system reliability basics. Some of the topics covered in this chapter are hardware reliability versus software reliability, major sources of computer failures, issues in computer system reliability, fault classifications, fault masking, and computer reliability measures. Chapter 5 is devoted to software reliability assessment and improvement methods. A number of methods grouped under seven categories are presented in this chapter.

Chapters 6 and 7 present various important aspects of software quality and human error and software bugs in computer systems, respectively. Chapter 8 is devoted to software safety and Internet reliability. It covers topics such as software safety classifications; potential software hazards; software safety assurance programs; software hazard analysis methods; Internet facts, figures, and examples; Internet outage classifications; and models for performing Internet reliability and availability analysis.

Chapter 9 covers various important aspects of software usability, including the need to consider usability during the software development phase, the software usability engineering process, software usability inspection methods, software usability test methods, and guidelines for conducting software usability testing. Chapter 10 is devoted to web usability. Some of the topics covered in the chapter are web usability facts and figures, common web design errors, web page design, tools for evaluating web usability, and questions to evaluate the effectiveness of website message communications. Finally, Chapter 11 presents various important aspects of computer system life-cycle costing.

This book will be useful to many individuals, including computer engineers, software engineers, design engineers, system engineers, human factors engineers, and other professionals involved with computers and the Internet—engineering managers and administrators; reliability and other engineers-at-large; researchers and instructors involved with computer systems; and graduate and senior undergraduate students in computer engineering, software engineering, system engineering, computer science, etc.

The author is deeply indebted to many individuals, including family members, friends, colleagues, and students, for their invisible input. The unseen contributions of my children also are appreciated. Last but not least, I thank my wife, Rosy, my other half and friend, for typing this entire book and her timely help in proofreading.

B. S. Dhillon
Ottawa, Ontario

About the Author

Dr. B. S. Dhillon is a professor of engineering management in the Department of Mechanical Engineering at the University of Ottawa. He has served as a chairman/director of the Mechanical Engineering Department/ Engineering Management Program for over ten years at the same institution. He is the founder of the probability distribution named the "Dhillon Distribution/Law/Model" by statistical researchers in their publications around the world. He has published over 364 (i.e., 217 [70 single authored and 147 coauthored] journal and 147 conference proceedings) articles on reliability engineering, maintainability, safety, engineering management, etc. He is or has been on the editorial boards of eleven international scientific journals. In addition, Dr. Dhillon has written forty books on various aspects of health care, engineering management, design, reliability, safety, and quality published by Wiley (1981), Van Nostrand (1982), Butterworth (1983), Marcel Dekker (1984), Pergamon (1986), etc. His books are being used in over one hundred countries, and many of them are translated into languages such as German, Russian, Chinese, and Persian (Iranian).

He has served as general chairman of two international conferences on reliability and quality control held in Los Angeles and Paris in 1987. Professor Dhillon has also served as a consultant to various organizations and bodies and has many years of experience in the industrial sector. At the University of Ottawa, he has been teaching reliability, quality, engineering management, design, and related areas for over thirty-three years, and he has also lectured in over fifty countries, including keynote addresses at various international scientific conferences held in North America, Europe, Asia, and Africa. In March 2004, Dr. Dhillon was a distinguished speaker at the Conference/Workshop on Surgical Errors (sponsored by White House Health and Safety Committee and Pentagon) held on Capitol Hill (One Constitution Avenue, Washington, DC).

Professor Dhillon attended the University of Wales, where he received a BS in electrical and electronic engineering and an MS in mechanical engineering. He received a PhD in industrial engineering from the University of Windsor.

1

Introduction

1.1 Background

Nowadays, computer systems have become an important element of the world economy, and each year billions of dollars are spent to develop, manufacture, operate, and maintain various types of computer systems around the globe. Their reliability, safety, and usability have become an important concern because of problems such as high cost, wrong decisions and actions, and accidental deaths. For example, a study conducted by the National Institute of Standards and Technology in 2002 found that software defects alone cost the U.S. economy about $59 billion annually, i.e., around 0.6% of its gross domestic product (GDP) [1].

The history of computer system reliability can be traced back to the late 1940s and 1950s to the works of Shannon [2], Hamming [3], Von Neumann [4], and Moore and Shannon [5]. For example, in 1956 Von Neumann [4] proposed the triple modular redundancy (TMR) scheme (nowadays widely used in computers) to improve system reliability. In 1965, Pierce published a book entitled *Failure Tolerant Design* [6]. This was probably the first book concerned with computer system reliability.

Over the years, a large number of publications related to computer system reliability have appeared. A comprehensive list of useful publications is provided in the appendix to this book.

1.2 Facts, Figures, and Examples

Some facts, figures, and examples concerned with computer system reliability are as follows:

- In 2002, a study commissioned by the National Institute of Standards and Technology (NIST) reported that software errors cost the U.S. economy about $59 billion annually [1].

- As per Kletz [7] and Herrman [8], the number of people killed due to computer system malfunctions worldwide up to the end of 1992 was somewhere between 1,000 and 3,000.
- The Internet has grown from four hosts in 1969 to over 147 million hosts and thirty-eight sites in 2002, and in 2001 there were over 52,000 Internet-related incidents and failures [9, 10].
- In 2000, the Internet economy generated about $830 billion in revenues in the United States [9–11].
- National Aeronautics and Space Administration's (NASA's) Saturn V Launch computer (circa 1964) had a mean time to failure (MTTF) of 25,000 hours [12–14].
- As per Landauer [15], an average software program contains about forty design flaws that impair the ability of workers to use it effectively.
- A pilot set the heading in a plane's computer-controlled inertial navigation system as 270° instead of 027°; the plane ran out of fuel and caused twelve fatalities [16].
- In 1966, European Space Agency's US$1 billion prototype Ariane 5 rocket was destroyed just forty seconds after launch because of a bug in the onboard guidance computer program [17].
- On April 25, 1997, a misconfigured router of a Virginia service provider injected an incorrect map into the global Internet. In turn, this caused network congestion, instability, and overload of Internet router table memory that ultimately shut down the majority of the main Internet backbones for about 2 hours [18].
- In 1963, a software error resulted in the incapacitation of a North American Air Defense Command (NORAD) exercise [19].
- As per Myers and Robson [20], 50% to 80% of all source code development accounts, directly or indirectly, for the user interface.
- A software error in the code controlling the Therac-25 radiation therapy machine caused many deaths in the 1980s [21].
- A computer opened the vent valve on the wrong vessel due to a software error, and fourteen tons of carbon dioxide were vented and lost [16, 22].
- As per Kuhn [23], a number of studies conducted over the years indicate that the reliability of Internet paths falls far short of the 99.999% availability expected in the public-switched telephone network (PSTN).
- In 1981, the launching of the first U.S. space shuttle was postponed for about twenty minutes prior to the scheduled launching time due to a software error [24].
- Some small-scale studies performed in 1994 and 2000 clearly reported that the probability of encountering a major routing pathology along a path (with respect to the Internet) was roughly 1.5% to 3.3% [25, 26].

- In 1991, a software fault caused a MIM-104 Patriot (surface-to-air missile system) to fail to intercept an incoming Iraqi Scud missile that caused twenty-eight American fatalities in Saudi Arabia [27].
- On August 14, 1998, a misconfigured Key Internet Database server mistakenly directed all queries for Internet machines with names ending in ".net" to the wrong secondary database server. This problem resulted in the failure of most connections to ".net" Internet web servers and other end stations for many hours [28].
- A case study performed in regard to Internet outages over a period of one year (November 1997–November 1998) categorized the outages under the following classifications (along with their occurrence percentages in parentheses) [28]:
  - Software problems (1.3%)
  - Malicious attacks (1.5%)
  - Sluggish/congestion (4.6%)
  - Unknown/undetermined/no problem (5.6%)
  - Miscellaneous (5.9%)
  - Routing problem (6.1%)
  - Interface down (6.2%)
  - Hardware problem (6.2%)
  - Unreachable (12.6%)
  - Fiber cut/circuit carrier problem (15.3%)
  - Power outage (16%)
  - Maintenance (16.2%)

1.3 Terms and Definitions

This section presents some useful terms and definitions concerned with various aspects of computer system reliability [14, 29–41].

Accident. An event that involves damage to a certain system/unit that suddenly disrupts the potential or current system/unit output.

Availability. The probability that an item/system is available for use or application when needed.

Debugging. The process of isolating and eradicating errors.

Downtime. The time period during which the item/system is not in a condition to perform its specified mission.

Failure. The inability of an item/system to perform its stated function.

Fault. An attribute that adversely affects the reliability of an item.

Fault-tolerant computing. The ability to execute specified algorithms successfully, irrespective of computer hardware malfunctions and software errors.

Hazard. The source of energy and the physiological and behavioral factors that, when uncontrolled, lead to harmful occurrences.

Human error. The failure to carry out a specified task (or the performance of a prohibited action) that could result in disruption of scheduled operations or damage to property.

Human factors. A body of scientific facts concerning the human characteristics (the term includes all biomedical and psychosocial considerations).

Mission time. The time during which the item/system is carrying out its stated mission.

Redundancy. The existence of more than one means to carry out a specified function.

Reliability. The probability that a system/unit/item will perform its assigned mission satisfactorily for the stated period of time when used according to the specified conditions.

Safety. Conservation of human life and the prevention of damage to systems/units/items as per mission-stated requirements.

Software error. A conceptual clerical or syntactic discrepancy that leads to one or more faults in the software.

Software reliability. The probability of a given software functioning for a specified time interval, without an error, when used according to the designed conditions on the stated machine.

Usability. The quality of an interactive system with respect to factors such as user satisfaction, ease of learning, and ease of use.

Usability evaluation. Any analytical or empirical study directed at assessing or understanding the usability of an interactive system/product.

User-centered design. An early and continuous involvement of potential users in the product design process.

User interface. The physical representations and procedures to view and interact with the product/system functionality.

1.4 Sources To Obtain Information Related to Computer System Reliability, Safety, and Usability

There are many different sources for obtaining information related to computer system reliability, safety, and usability. These include journals and magazines, conference proceedings, industry standards and handbooks, data sources, and books. Some of these sources are listed in the following subsections [14, 29, 30]:

1.4.1 Journals and Magazines

ACM Transactions on Computer-Human Interaction (TOCHI)
Human-Computer Interaction
IEEE Transactions on Reliability
Interacting with Computers
International Journal of Industrial Ergonomics
International Journal of Man-Machine Studies
International Journal of Reliability, Quality, and Safety Engineering
Journal of Safety Research
National Safety News
Professional Safety
Quality and Reliability Engineering
Reliability Engineering and System Safety
Reliability Review
Safety Science
User Modeling and User-Adopted Interaction (UMUAI)

1.4.2 Conference Proceedings

Proceedings of the Annual Conference on Computer Assurance (USA)

Proceedings of the Annual Reliability and Maintainability Symposium (USA)

Proceedings of the Annual Reliability Engineering Conference for the Electric Power Industry (USA)

Proceedings of the IEEE International Conference on Systems, Man, and Cybernetics (USA)

Proceedings of the International Conference on Reliability and Exploitation of Computer Systems (Poland)

Proceedings of the Symposium on Reliability in Electronics (Hungary)

1.4.3 Standards and Handbooks

ANSI/AIAA R-103, *Recommended Practice for Software Reliability*, American National Standards Institute (ANSI), New York.

ETSI ETR 095, *Human Factors: Guide for Usability Evaluations of Telecommunications Systems and Services,* European Telecommunications Standards Institute (ETSI), Sophia Antipolis, France.

ISO 9241-11 (1988), *Ergonomics Requirements for Office Work with Visual Display Terminals (VDTs): Guidance on Usability,* International Organization for Standardization (ISO), Geneva, Switzerland.

ISO 9241-13 (1998), *Ergonomics Requirements for Office Work with Visual Display Terminals (VDTs): User Guidance,* International Organization for Standardization (ISO), Geneva, Switzerland.

ISO 13407 (1999), *Human Centered Design Processes for Interactive Systems,* International Organization for Standardization (ISO), Geneva, Switzerland.

MIL-HDBK-217, *Reliability Prediction of Electronic Equipment,* Department of Defense, Washington, DC.

MIL-HDBK-338, *Electronics Reliability Design Handbook,* Department of Defense, Washington, DC.

MIL-STD-781, *Reliability Design, Qualification and Production Acceptance Tests: Exponential Distribution,* Department of Defense, Washington, DC.

MIL-STD-790, *Reliability Assurance Program for Electronic Parts Specifications,* Department of Defense, Washington, DC.

MIL-STD-1629, *Procedures for Performing a Failure Mode, Effects and Criticality Analysis,* Department of Defense, Washington, DC.

MIL-STD-1908, *Definitions of Human Factors Terms,* Department of Defense, Washington, DC.

MIL-STD-2167, *Defense System Software Development,* Department of Defense, Washington, DC.

MIL-STD-52779, *Software Quality Assurance Program Requirements,* Department of Defense, Washington, DC.

1.4.4 Data Sources

American National Standards Institute, 11 W. 42nd St., New York, NY 10036.

Computer Accident/Incident Report System, System Safety Development Center, EG&G, P.O. Box 1625, Idaho Falls, ID.

Gertman, D. I., and Blackman, H. S., *Human Reliability and Safety Analysis Data Handbook,* John Wiley and Sons, New York, 1994.

Government Industry Data Exchange Program (GIDEP), GIDEP Operations Center, U.S. Department of Navy, Naval Weapons Station, Seal Beach, Corona, CA 91720.

National Aeronautics and Space Administration (NASA), Parts Reliability Information Center, George C. Marshall Space Flight Center, Huntsville, AL 35812.

National Technical Information Center (NTIS), 5285 Port Royal Road, Springfield, VA 22151.

Reliability Analysis Center, Rome Air Development Center (RADC), Griffiss Air Force Base, Rome, NY 13441-5700.

Safety Research Information Service, National Safety Council, 444 North Michigan Avenue, Chicago, IL.

System Reliability Service, Safety and Reliability Directorate, UKAEA, Wigshaw Lane, Culcheth, Warrington WA3 4NE, UK.

1.4.5 Books

Bawa, J., Dorazio, P., and Trenner, L., eds., *The Usability Business: Making the Web Work*, Springer-Verlag, New York, 2001.

Brink, T., Gergle, D., and Wood, S. D., *Designing Websites that Work: Usability for the Web*, Morgan Kaufmann Publishers, San Francisco, 2002.

Coderre, D., *Computer-Aided Fraud Prevention and Detection: A Step-by-Step Guide*, John Wiley and Sons, New York, 2009.

Dhillon, B. S., *Human Reliability: With Human Factors*, Pergamon Press, New York, 1986.

Dhillon, B. S., *Reliability in Computer System Design*, Ablex Publishing Corporation, Norwood, NJ, 1987.

Dhillon, B. S., *Engineering Safety: Fundamentals, Techniques, and Applications*, World Scientific Publishing, River Edge, NJ, 2003.

Dhillon, B. S., *Engineering Usability: Fundamentals, Applications, Human Factors, and Human Error*, American Scientific Publishers, Stevenson Ranch, CA, 2004.

Glass, R. L., *Software Reliability Guidebook*, Prentice Hall, Englewood Cliffs, NJ, 1979.

Leveson, N. G., *Safeware: System Safety and Computers*, Addison-Wesley, Reading, MA, 1995.

Lindgaard, G., *Usability Testing and System Evaluation: A Guide for Designing Useful Computer Systems*, Chapman and Hall, London, 1994.

Longbottom, R., *Computer Systems Reliability*, John Wiley and Sons, New York, 1980.

Pierce, W. H., *Failure-Tolerant Computer Design*, Academic Press, New York, 1965.

Rosson, M. B., and Carroll, J. M., *Usability Engineering: Scenario-Based Development of Human-Computer Interaction*, Academic Press, San Francisco, 2002.

Sanders, M. S., and McCormick, E. J., *Human Factors in Engineering and Design*, McGraw Hill, New York, 1993.

Spellman F. R., and Whiting, N. E., *Safety Engineering: Principles and Practice*, Government Institutes, Rockville, MD, 1999.

Stephans, R. A., and Talso, W. W., eds., *System Safety Analysis Handbook*, System Safety Society, Irvine, CA, 1993.

Wilcox, R. H., *Redundancy Techniques for Computing Systems*, Spartan Books, New York, 1962.

1.5 Scope of the Book

Computers have become an important element of world economy because each year a vast sum of money is spent to develop, manufacture, operate, and maintain various types of computer systems around the globe. Their reliability, safety, and usability have become an important concern because of problems such as high cost, wrong decisions and actions, and accidental deaths. Over the years, a large number of publications concerned with computer system reliability, safety, and usability have appeared. At present, to the best of the author's knowledge, there is no specific book on the topic. This book not only attempts to cover computer system reliability, safety, and usability within its framework, but also provides the latest developments in the area.

Finally, the main objective of this book is to provide professionals and others concerned with computer systems with information that could be useful to improve these systems' reliability, safety, and usability. This book will be useful to many individuals, including computer engineers, software engineers, design engineers, system engineers, human factors engineers, and other professionals involved with computers and the Internet—engineering managers and administrators, reliability engineers and other at-large engineers, researchers and instructors involved with computer systems, and graduate and senior undergraduate students in computer engineering, software engineering, system engineering, computer science, etc.

Problems

1. Write an essay on computer system reliability.
2. List at least five important facts and figures concerned with computer system reliability.

3. Define the following terms:
   a. Fault-tolerant computing
   b. Software reliability
   c. Software error
4. List at least eight journals considered most useful to obtain computer system reliability-related information.
5. List at least six useful sources for obtaining computer system reliability-related data.
6. Compare computer system hardware reliability with software reliability.
7. Define the following terms:
   a. Safety
   b. Usability
   c. Reliability
8. List at least five standards considered most useful with respect to computer system reliability.
9. List and discuss at least six books considered most useful to obtain directly or indirectly computer system reliability-related information.
10. Compare computer system reliability with computer system availability.

___

References

1. National Institute of Standards and Technology (NIST). (2002). 100 Bureau Drive, Stop 1070, Gaithersburg, MD.
2. Shannon, C. E. (1948). A mathematical theory of communications. *Bell System Tech. Journal, 27*, 379–423, 623–656.
3. Hamming, W. R. (1950). Error detecting and error correcting codes. *Bell System Tech. Journal, 29*, 147–160.
4. Von Neumann, J. (1956). Probabilistic logics and the synthesis of reliable organisms from reliable components. In C. E. Shannon & J. McCarthy (Eds.), *Automata studies* (pp. 43–48). Princeton, NJ: Princeton University Press.
5. Moore, E. F., & Shannon, C. E. (1956). Reliable circuits using less reliable relays. *Journal of Franklin Institute, 262*, 191–208.
6. Pierce, W. H. (1965). *Failure-tolerant computer design.* New York, NY: Academic Press.
7. Kletz, T. (1997). Reflections on safety. *Safety Systems, 6*(3), 1–3.
8. Herrman, D. S. (1999). *Software safety and reliability.* Los Alamitos, CA: IEEE Computer Society Press.
9. Hafner, K., & Lyon, M. (1996). *Where wizards stay up late: The origins of the Internet.* New York, NY: Simon and Schuster.

10. Dhillon, B. S. (2007). *Applied reliability and quality: Fundamentals, methods, and procedures*. London: Springer.
11. Goseva-Popstojanova, K., Mazimdar, S., & Singh, A. D. (2004). Empirical study of session-based workload and reliability for web servers. *Proceedings of the 15th International Symposium on Software Reliability Engineering*, 403–414.
12. Pradhan, D. K. (Ed.). (1986). *Fault-tolerant computing theory and techniques* (Vols. 1 and 2). Englewood Cliffs, NJ: Prentice Hall.
13. Shooman, M. L. (1994). Fault-tolerant computing. *Annual Reliability and Maintainability Symposium Tutorial Notes*, 1–25.
14. Dhillon, B. S. (1999). *Design reliability: Fundamentals and applications*. Boca Raton, FL: CRC Press.
15. Landauer, T. (1995). *The trouble with computers*. Boston: Massachusetts Institute of Technology Press.
16. Kletz, T. (with Chung, P., Broomfield, E., & Shen-Orr, C.). (1995). *Computer control and human error*. Houston, TX: Gulf Publishing.
17. Dowson, M. (1997). The Ariane 5 software failure. *Software Engineering Notes, 22*(2), 84–85.
18. Barrett, R., Haar, S., & Whitestone, R. (1997, April 25). Routing snafu causes Internet outage. *Interactive Week*, p. 9.
19. Myers, G. J. (1976). *Software reliability principles and practices*. New York, NY: John Wiley and Sons.
20. Myers, B., & Robson, M. B. (1992). Survey on user interface programming. *Proceedings of the ACM CHI'92 Human Factors in Computing Systems Conference*, 195–202.
21. Leveson, N., & Turner, C. S. (1993). An investigation of the Therac-25 accidents. *IEEE Trans. on Computers, 26*(7), 18–41.
22. Nimmo, I. (1994). Extend HAZOP to computer control systems. *Chemical Engineering Progress, 90*(10), 32–44.
23. Kuhn, D. R. (1997). Sources of failure in the public switched telephone networks. *IEEE Trans. on Computers, 30*(4), 31–36.
24. Graman, J. R. (1981). The bug heard around the world. *Software Engineering Notes, 6*(October), 3–10.
25. Gummadi, K. P., Madhyastha, H. V., Gribble, S. D., Levy, H. M., & Wetherall, D. (2004). Improving the reliability of Internet paths with one-hop source routing. *Proceedings of the 6th Usenix/ACM Symposium on Operating Systems and Design Implementation (OSDI)*, 183–198.
26. Paxson, V. (1997). End-to-end routing behavior in the Internet. *IEEE/ACM Transactions on Networking, 5*(5), 278–285.
27. United States Government Accounting Office (GAO). (1992). *Patriot missile defense, software problem led to system failure at Dharhan, Saudi Arabia* (Report No. IMTEC 92-26). Washington, DC: Author.
28. Lapovitz, C., Ahuja, A., & Jahamian, F. (1999). Experimental study of Internet stability and wide-area backbone failures. *Proceedings of the 29th Annual International Symposium on Fault-Tolerant Computing*, 601–615.
29. Dhillon, B. S. (2004). *Engineering usability: Fundamentals, applications, human factors, and human error*. Stevenson Ranch, CA: American Scientific Publishers.
30. Dhillon, B. S. (2013). *Safety and human error in engineering systems*. Boca Raton, FL: CRC Press.

31. Rosson, M. B., & Carroll, J. M. (2002). *Usability engineering: Scenario-based development of human-computer interaction*. San Francisco, CA: Academic Press.
32. International Organization for Standardization (ISO). (1999). *User-centered design process for interactive systems* (ISO 13407). Geneva, Switzerland: Author.
33. Department of Defense. (1986). *Definitions of effectiveness terms for reliability, maintainability, human factors, and safety* (MIL-STD-721B). Washington, DC: Author.
34. Meister, D. (1966). Human factors in reliability. In W. G. Ireson (Ed.), *Reliability handbook* (pp. 12.2–12.37). New York, NY: McGraw Hill.
35. *Dictionary of terms used in the safety profession*. (1988). Des Plaines, IL: American Society of Safety Engineers.
36. Meulen, M. V. D. (2000). *Definitions for hardware and software engineers*. London: Springer-Verlag.
37. Institute of Electrical and Electronic Engineers (IEEE). (1994). *Standard for software safety plans* (IEEE-STD-1228). New York, NY: Author.
38. Dhillon, B. S. (1986). *Human reliability: With human factors*. New York, NY: Pergamon Press.
39. Department of Defense. (1992). *Definitions of human factors terms* (MIL-STD-1908). Washington, DC: Author.
40. Naresky, J. J. (1970). Reliability definitions. *IEEE Trans. on Reliability, 19*, 198–200.
41. Omdahl, T. P. (Ed.). (1988). *Reliability, availability, and maintainability (RAM) dictionary*. Milwaukee, WI: ASQC Quality Press.

2

Basic Mathematical Concepts

2.1 Introduction

As in the development of other areas of science and technology, mathematics has also played a pivotal role in the development of computer system reliability and its associated areas. The history of mathematics may be traced back to the development of our current number symbols, sometimes referred to as the "Hindu-Arabic numeral system" in the published literature [1]. The very first evidence of the use of these symbols is found on stone columns erected by the Scythian emperor Asoka of India in around 250 BCE [1].

However, the thinking on the concept of probability is relatively new, tracing back to a gambler's manual written by Girolamo Cardano (1501–1576) [2], in which he considered some interesting probability-related issues. However, Pierre Fermat (1601–1665) and Blaise Pascal (1623–1662) were the first two persons who independently and correctly solved the problem of dividing the winnings in a game of chance.

Laplace transforms, often used to find solutions to differential equations, were developed by Pierre-Simon Laplace (1749–1827). Needless to say, additional information on the historical developments in mathematics is available in the literature [1, 2]. This chapter presents some mathematical concepts that will be helpful in understanding subsequent chapters of this book.

2.2 Arithmetic Mean and Mean Deviation

A set of data is only useful if it is analyzed properly. Certain characteristics are important in describing the nature of a given data set. This section presents two statistical measures that are useful in performing computer system reliability-related analysis [3–5].

2.2.1 Arithmetic Mean

Generally, this is simply referred to as the mean and is expressed by

$$m = \frac{\sum_{i=1}^{k} x_i}{k} \tag{2.1}$$

where
   m = the mean
   k = number of values
   x_i = value i for $i = 1,2,3,\ldots,k$

Example 2.1

Assume that the inspection department of a computer manufacturer inspected eight identical computers and found 4, 8, 5, 3, 9, 6, 3, and 2 defects in each respective computer. Calculate the average number of defects per computer.

By substituting the given data values into Equation (2.1), we get:

$$m = \frac{4+8+5+3+9+6+3+2}{8}$$

$$= 5 \text{ defects per computer}$$

Thus, the average number of defects per computer is 5, i.e., the arithmetic mean of the given data set is 5 defects per computer.

2.2.2 Mean Deviation

This is a measure of dispersion and is expressed by

$$D_m = \frac{\sum_{j=1}^{k} |x_j - m|}{k} \tag{2.2}$$

where
   D_m = the mean deviation
   k = number of data values
   x_j = data value j for $j = 1,2,3,\ldots,k$
   m = the mean of the given data set values
   $|x_j - m|$ = absolute value of the deviation of x_j from m

Example 2.2

Calculate the mean deviation of the set of data values given in Example 2.1.

In Example 2.1, the calculated mean value of the given data set is 5 defects per computer. Thus, by substituting this calculated value and the given data values into Equation (2.2), we get

$$D_m = \frac{\left[|4-5|+|8-5|+|5-5|+|3-5|+|9-5|+|6-5|+|3-5|+|2-5|\right]}{8}$$

$$= \frac{1+3+0+2+4+1+2+3}{8}$$

$$= 2$$

Thus, the mean deviation of the given data set is 2.

2.3 Boolean Algebra Laws

Boolean algebra is named after George Boole (1815–1864), and it plays an important role in probability theory and reliability and safety-related studies. Some of its laws are useful in understanding subsequent chapters of this book [6–8].

- Commutative law

$$A + B = B + A \tag{2.3}$$

$$A.B = B.A \tag{2.4}$$

where
A = an arbitrary set or event
B = an arbitrary set or event
Dot (.) = the intersection of sets (Note that sometimes Equation [2.4] is written without the dot, but it still conveys the same meaning.)
+ = the union of sets
- Idempotent law

$$A + A = A \tag{2.5}$$

$$AA = A \tag{2.6}$$

- Absorption law

$$A + (AB) = A \tag{2.7}$$

$$B(B + A) = B \tag{2.8}$$

- Associative law

$$(AB)C = A(BC) \tag{2.9}$$

$$(A + B) + C = A + (B + C) \tag{2.10}$$

where
   C = an arbitrary set or event
- Distributive law

$$A(B + C) = AB + AC \tag{2.11}$$

$$(A + B)(A + C) = A + BC \tag{2.12}$$

2.4 Probability Definition and Properties

Probability may be defined as follows [9]:

$$P(X) = \lim_{n \to \infty}(N/n) \tag{2.13}$$

where
   $P(X)$ = probability of occurrence of event X
   N = number of times event X occurs in the n repeated experiments

The following list presents some of the basic properties of probability [8, 9]:

- The probability of occurrence of event X is

$$0 \le P(X) \le 1 \tag{2.14}$$

- The probability of occurrence and nonoccurrence of event X is always

$$P(X) + P(\bar{X}) = 1 \tag{2.15}$$

where
   $P(X)$ = occurrence probability of event X
   $P(\bar{X})$ = nonoccurrence probability of event X
- The probability of the union of n mutually exclusive events is

$$P(Y_1 + Y_2 + \ldots + Y_n) = \sum_{i=1}^{n} P(Y_i) \tag{2.16}$$

where
   $P(Y_i)$ = probability of occurrence of event Y_i for $i = 1,2,3,\ldots,n$

- The probability of the union of n independent events is

$$P(Y_1 + Y_2 + \ldots + Y_n) = 1 - \prod_{i=1}^{n}\left(1 - P(Y_i)\right) \tag{2.17}$$

- The probability of an intersection of n independent events is

$$P(Y_1 Y_2 Y_3 \ldots Y_n) = P(Y_1)\,P(Y_2)\,P(Y_3) \ldots P(Y_n) \tag{2.18}$$

Example 2.3

A computer system has two critical subsystems Y_1 and Y_2. The failure of either subsystem can result in a serious accident. The probabilities of failure of subsystems Y_1 and Y_2 are 0.04 and 0.06, respectively. Calculate the probability of the occurrence of a serious accident if both of these subsystems fail independently.

By substituting the given data values into Equation (2.17), we get

$$P(Y_1 + Y_2) = 1 - \prod_{i=1}^{2}\left(1 - P(Y_i)\right)$$

$$= P(Y_1) + P(Y_2) - [P(Y_1)P(Y_2)]$$

$$= 0.04 + 0.06 - [(0.04)(0.06)]$$

$$= 0.0976$$

Thus, the probability of the occurrence of a serious accident is 0.0976.

2.5 Probability-Related Definitions

This section presents a number of probability-related definitions that will be useful in understanding subsequent chapters of this book.

2.5.1 Cumulative Distribution Function

For a continuous random variable, the cumulative distribution function is defined by [9]

$$F(t) = \int_{-\infty}^{t} f(y)dy \tag{2.19}$$

where
y = a continuous random variable
$F(t)$ = cumulative distribution function
$f(y)$ = probability density function

For $t = \infty$, Equation (2.19) yields

$$F(\infty) = \int_{-\infty}^{\infty} f(y)\,dy$$

$$= 1 \tag{2.20}$$

It simply means that the total area under the probability curve is equal to unity.

Example 2.4

Assume that the failure (probability) density function of a computer system is expressed by

$$f(t) = \lambda e^{-\lambda t}, \quad \text{for } t \ge 0, \lambda > 0 \tag{2.21}$$

where
t = time (i.e., a continuous random variable)
f(t) = probability density function (In the area of reliability engineering, it is called the *failure density function*.)
λ = computer system failure rate

Obtain an expression for the computer system cumulative distribution function.
By substituting Equation (2.21) into Equation (2.19), we obtain

$$F(t) = \int_{0}^{t} \lambda e^{-\lambda t}\,dt \tag{2.22}$$

$$= 1 - e^{-\lambda t}$$

Thus, Equation (2.22) is the expression for the computer system cumulative distribution function.

2.5.2 Probability Density Function

For a continuous random variable, the probability density function is expressed by [9, 10]

$$f(t) = \frac{dF(t)}{dt} \tag{2.23}$$

where
f(t) = probability density function
F(t) = cumulative distribution function

Example 2.5

Prove with the aid of Equation (2.22) that Equation (2.21) is the probability density function.

By inserting Equation (2.22) into Equation (2.23), we obtain

$$f(t) = \frac{d\left(1 - e^{-\lambda t}\right)}{dt}$$

$$= \lambda e^{-\lambda t} \tag{2.24}$$

Equations (2.21) and (2.24) are identical.

2.5.3 Expected Value

The expected value of a continuous random variable is expressed by [9]

$$E(t) = \int_{-\infty}^{\infty} tf(t)\,dt \tag{2.25}$$

where
$E(t)$ = the expected value (or mean value) of the continuous random variable t

Example 2.6

Find the expected value or mean value of the failure (probability) density function expressed by Equation (2.21).
   By inserting Equation (2.21) into Equation (2.25), we obtain

$$E(t) = \int_{0}^{\infty} t\lambda e^{-\lambda t}\,dt$$

$$= \left[-te^{-\lambda t}\right]_{0}^{\infty} - \left[-\frac{e^{-\lambda t}}{\lambda}\right]_{0}^{\infty} \tag{2.26}$$

$$= \frac{1}{\lambda}$$

Thus, the expected value or mean value of the failure (probability) density function expressed by Equation (2.21) is given by Equation (2.26).

2.6 Probability Distributions

There are a large number of probability distributions in the published literature [11, 12]. This section presents four probability distributions considered useful in performing computer system reliability-related analysis.

2.6.1 Exponential Distribution

This is a widely used probability distribution to perform various types of reliability analysis. Its probability density function is defined by [12, 13]

$$f(t) = \lambda e^{-\lambda t}, \quad \text{for } t \geq 0, \lambda > 0 \tag{2.27}$$

where
  $f(t)$ = probability density function
  t = time
  λ = distribution parameter (In the area of reliability engineering, it is often referred to as unit/part/system failure rate.)

By inserting Equation (2.27) into Equation (2.19), we get

$$F(t) = \int_0^t \lambda e^{-\lambda t} \, dt$$
$$= 1 - e^{-\lambda t} \tag{2.28}$$

where
  $F(t)$ = cumulative distribution function

By substituting Equation (2.27) into Equation (2.25), we obtain

$$E(t) = \int_0^\infty t \lambda e^{-\lambda t} \, dt$$
$$= \frac{1}{\lambda} \tag{2.29}$$

where
  $E(t)$ = the expected value or mean value of the exponential distribution

Example 2.7

Assume that the failure rate of a computer system is 0.0004 failures per hour. Calculate the probability of a failure occurrence during a 500-hour mission by using Equation (2.28). Thus, we have t = 500 hours and λ = 0.0004 failures per hour.
  By substituting these two values into Equation (2.28), we get

$$F(500) = 1 - e^{-(0.0004)(500)}$$

$$= 0.1813$$

This means that there is an 18.13% chance that a failure will occur during the 500-hour mission.

2.6.2 Rayleigh Distribution

This probability distribution is named after its founder, John Rayleigh (1842–1919) [1]. The distribution probability density function is defined by

$$f(t) = \frac{2te^{-\left(\frac{t}{\alpha}\right)^2}}{\alpha^2}, \quad \text{for } t \geq 0, \alpha > 0 \tag{2.30}$$

where
  α = the distribution parameter

By substituting Equation (2.30) into Equation (2.19), we obtain the following equation for the cumulative distribution function:

$$F(t) = 1 - e^{-\left(\frac{t}{\alpha}\right)^2} \tag{2.31}$$

Inserting Equation (2.30) into Equation (2.25) yields the following expression for the expected value of t:

$$E(t) = \alpha \Gamma (3/2) \tag{2.32}$$

where
  $\Gamma(.)$ = the gamma function, which is expressed by

$$\Gamma(n) = \int_0^\infty t^{n-1}e^{-t}\,dt, \quad \text{for } n > 0 \tag{2.33}$$

2.6.3 Weibull Distribution

This probability distribution is named after its founder, W. Weibull, who developed it in the early 1950s [14]. The distribution probability density function is defined by

$$f(t) = \frac{bt^{b-1}}{\alpha^b}e^{-\left(\frac{t}{\alpha}\right)^b}, \quad \text{for } t \geq 0, \alpha > 0, b > 0 \tag{2.34}$$

where
  b and α = the shape and scale parameters, respectively

Substituting Equation (2.34) into Equation (2.19) yields the following equation for the cumulative distribution function:

$$F(t) = 1 - e^{-\left(\frac{t}{\alpha}\right)^b} \tag{2.35}$$

By inserting Equation (2.34) into Equation (2.25), we get the following equation for the expected value of *t*:

$$E(t) = \alpha\Gamma\left(1+\frac{1}{b}\right) \tag{2.36}$$

Note that for $b = 1$ and $b = 2$, the exponential and Rayleigh distributions are the special cases of this probability distribution, respectively.

2.6.4 Bathtub Hazard Rate Curve Distribution

This probability distribution was developed in 1981 [15], and in the published literature by other authors, it is generally referred to as the Dhillon distribution/law/model [16–34]. The distribution can represent decreasing, increasing, and bathtub-shape hazard rates.

The distribution probability density function is defined by [15]

$$f(t) = b\theta(\theta t)^{b-1} e^{\left\{e^{(\theta t)^b} - (\theta t)^b - 1\right\}}, \quad \text{for } b > 0, \theta > 0, t \geq 0 \tag{2.37}$$

where

θ and b = the scale and shape parameters, respectively

By substituting Equation (2.37) into Equation (2.19), we obtain the following equation for the cumulative distribution function:

$$F(t) = 1 - e^{-\left\{e^{(\theta t)^b} - 1\right\}} \tag{2.38}$$

Note that for $b = 0.5$, this distribution gives the bathtub hazard rate curve, and for $b = 1$ it gives the extreme value distribution. In other words, the extreme value distribution is the special case of this distribution at $b = 1$.

2.7 Laplace Transform Definition, Common Laplace Transforms, and Final-Value Theorem's Laplace Transform

The Laplace transform (named after Pierre-Simon Laplace [1749–1827]) of a function, say $f(t)$, is defined by

$$F(s) = \int_0^\infty f(t)e^{-st}\,dt \tag{2.39}$$

TABLE 2.1

Laplace Transforms of Some Functions

No.	f(t)	F(s)
1	C (a constant)	$\dfrac{c}{s}$
2	$e^{-\lambda t}$	$\dfrac{1}{s+\lambda}$
3	t	$\dfrac{1}{s^2}$
4	$te^{-\lambda t}$	$\dfrac{1}{(s+\lambda)^2}$
5	$\lambda_1 f_1(t) + \lambda_2 f_2(t)$	$\lambda_1 F_1(s) + \lambda_2 F_2(s)$
6	$\dfrac{df(t)}{dt}$	$sF(s) - f(0)$

where

$F(s)$ = Laplace transform of $f(t)$

s = Laplace transform variable

t = a variable

Laplace transforms of some commonly occurring functions in the computer system reliability-related studies are presented in Table 2.1 [35, 36].

2.7.1 Laplace Transform: Final-Value Theorem

If the following limits exist, then:

$$\lim_{t \to \infty} f(t) = \lim_{s \to 0} sF(s) \qquad (2.40)$$

Example 2.8

Prove with the aid of the following equation that the left side of Equation (2.40) is equal to its right side:

$$f(t) = \frac{\theta}{(\lambda+\theta)} + \frac{\lambda}{(\lambda+\theta)} e^{-(\lambda+\theta)t} \qquad (2.41)$$

where

λ and μ = constants

By inserting Equation (2.41) into the left side of Equation (2.40), we obtain

$$\lim_{t \to \infty} \left[\frac{\theta}{(\lambda+\theta)} + \frac{\lambda}{(\lambda+\theta)} e^{-(\lambda+\theta)t} \right] = \frac{\theta}{\lambda+\theta} \qquad (2.42)$$

With the aid of Table 2.1, we get the following Laplace transforms of Equation (2.41):

$$F(s) = \frac{\theta}{s(\lambda+\theta)} + \frac{\lambda}{(\lambda+\theta)} \cdot \frac{1}{(s+\lambda+\theta)} \tag{2.43}$$

By substituting Equation (2.43) into the right side of Equation (2.40), we get

$$\lim_{s\to 0} s\left[\frac{\theta}{s(\lambda+\theta)} + \frac{\lambda}{(\lambda+\theta)} \cdot \frac{1}{(s+\lambda+\theta)}\right] = \frac{\theta}{\lambda+\theta} \tag{2.44}$$

As the right sides of both Equations (2.42) and (2.44) are exactly the same, it proves that the left side of Equation (2.40) is equal to its right side.

2.8 Laplace Transforms' Application in Solving First-Order Differential Equations

Laplace transforms are often used to find solution to first-order linear differential equations in computer system reliability-related studies. The application of Laplace transforms to find a solution to a set of linear differential equations describing a computer system with respect to reliability is demonstrated through the following example.

Example 2.9

Assume that a computer system can be in one of two states: operating normally or failed. The following two first-order linear differential equations describe the system:

$$\frac{dP_{on}(t)}{dt} + \lambda_c P_{on}(t) = 0 \tag{2.45}$$

$$\frac{dP_f(t)}{dt} - \lambda_c P_{on}(t) = 0 \tag{2.46}$$

where
$P_j(t)$ = probability that the computer system is in state j at time t for j = on (operating normally) and j = f (failed)
λ_c = computer system constant failure rate

And at time $t = 0$, $P_{on}(0) = 1$ and $P_f(0) = 0$.
   Find solutions to Equations (2.45) and (2.46) with the aid of Laplace transforms.

Taking Laplace transforms of Equations (2.45) and (2.46) and using the given initial conditions yields

$$P_{on}(s) = \frac{1}{s + \lambda_c}$$

(2.47)

$$P_f(s) = \frac{\lambda_c}{s(s + \lambda_c)}$$

(2.48)

where
$P_j(s)$ = Laplace transform of the probability that the computer system is in state j at time t for j = on (operating normally) or j = f (failed)

The inverse Laplace transforms of Equations (2.47) and (2.48) are

$$P_{on}(t) = e^{-\lambda_c t}$$

(2.49)

and

$$P_f(t) = 1 - e^{-\lambda_c t}$$

(2.50)

Thus, Equations (2.49) and (2.50) are the solutions to Equations (2.45) and (2.46).

Problems

1. Mathematically define mean deviation.
2. The inspection department of a computer manufacturer inspected five identical computers and found 7, 20, 25, 15, and 4 defects in each computer. Calculate the average number of defects per computer.
3. What is idempotent law?
4. Prove that the left side of Equation (2.12) is equal to its right side.
5. What are the basic properties of probability?
6. What is the difference between mutually and nonmutually exclusive events?
7. Define cumulative distribution function.
8. What are the special-case distributions of the Weibull probability distribution?
9. Write down the probability density function for the bathtub hazard rate curve distribution.
10. A computer system has two critical subsystems, say, X1 and X2. The failure of either subsystem can lead to a serious accident. The probabilities of failure of subsystems X1 and X2 are 0.09 and 0.05, respectively. Calculate the probability of the occurrence of a serious accident if both subsystems fail independently.

References

1. Eves, H. (1976). *An introduction to the history of mathematics.* New York, NY: Holt, Rinehart, Winston.
2. Owen, D. B. (Ed.). (1976). *History of statistics and probability.* New York, NY: Marcel Dekker.
3. Speigel, M. R. (1961). *Statistics.* New York, NY: McGraw Hill.
4. Dhillon, B. S. (2004). *Reliability, quality, and safety for engineers.* Boca Raton, FL: CRC Press.
5. Spiegel, M. R. (1975). *Probability and statistics.* New York, NY: McGraw Hill.
6. Nuclear Regulatory Commission. (1981). *Fault tree handbook* (Report No. NUREG-0492). Washington, DC: Author.
7. Lipschutz, S. (1964). *Set theory.* New York, NY: McGraw Hill.
8. Lipschutz, S. (1965). *Probability.* New York, NY: McGraw Hill.
9. Mann, N. R., Schafer, R. E., & Singpurwalla, N. D. (1974). *Methods for statistical analysis of reliability and life data.* New York, NY: John Wiley and Sons.
10. Shooman, M. L. (1968). *Probabilistic reliability: An engineering approach.* New York, NY: McGraw Hill.
11. Patel, J. K., Kapadia, C. H., & Owen, D. H. (1976). *Handbook of statistical distributions.* New York, NY: Marcel Dekker.
12. Dhillon, B. S. (1983). Reliability engineering. In *Systems design and operation.* New York, NY: Van Nostrand Reinhold.
13. Davis, D. J. (1952). Analysis of some failure data. *J. Amer. Stat. Assoc., 113–150.*
14. Weibull, W. (1951). A statistical distribution function of wide applicability. *J. Appl. Mech., 18,* 293–297.
15. Dhillon, B. S. (1981). Life distributions. *IEEE Transactions on Reliability, 30,* 457–460.
16. Baker, R. D. (1993). Nonparametric estimation of the renewal function. *Computers Operations Research, 20*(2), 167–178.
17. Cabana, A., & Cabana, E. M. (2005). Goodness-of-fit to the exponential distribution, focused on Weibull alternatives. *Communications in Statistics-Simulation and Computation, 34,* 711–723.
18. Grane, A., & Fortiana, J. (2011). A directional test of exponentiality based on maximum correlations. *Metrika, 73,* 255–274.
19. Henze, N., & Meintnis, S. G. (2005). Recent and classical tests for exponentiality: A partial review with comparisons. *Metrika, 61,* 29–45.
20. Hollander, M., Laird, G., & Song, K. S. (2003). Nonparametric interference for the proportionality function in the random censorship model. *Nonparametric Statistics, 15*(2), 151–169.
21. Jammalamadaka, S. R., & Taufer, E. (2003). Testing exponentiality by comparing the empirical distribution function of the normalized spacings with that of the original data. *Journal of Nonparametric Statistics, 15*(6), 719–729.
22. Jammalamadaka, S. R., & Taufer, E. (2006). Use of mean residual life in testing departures from exponentiality. *Journal of Nonparametric Statistics, 18*(3), 277–292.
23. Kunitz, H. (1989). A new class of bathtub-shaped hazard rates and its application in comparison of two test-statistics. *IEEE Transactions on Reliability, 38*(3), 351–354.

24. Kunitz, H., & Pamme, H. (1993). The mixed gamma aging model in life data analysis. *Statistical Papers, 34,* 303–318.
25. Meintanis, S. G. (2009). A class of tests for exponentiality based on a continuum of moment conditions. *Kybernetika, 45*(6), 946–959.
26. Morris, K., & Szynal, D. (2008). Some U-statistics in goodness-of-fit tests derived from characterizations via record values. *International Journal of Pure and Applied Mathematics, 46*(4), 339–414.
27. Morris, K., & Szynal, D. (2007). Goodness-of-fit tests based on characterizations involving moments of order statistics. *International Journal of Pure and Applied Mathematics, 38*(1), 83–121.
28. Na., M. H. (1999). Spline hazard rate estimation using censored data. *Journal of KSIAM, 3*(2), 99–106.
29. Nam, K. H., & Chang, S. J. (2006). Approximation of the renewal function for Hjorth model and Dhillon model. *Journal of the Korean Society for Quality Management, 34*(1), 34–39.
30. Nam, K. H., & Park, D. H. (1997). Failure rate for Dhillon model. *Proceedings of the Spring Conference of the Korean Statistical Society*, 114–118.
31. Nimoto, N., & Zitikis, R. (2008). The Atkinson index, the Moran statistic, and testing exponentiality. *Journal of the Japan Statistical Society, 38*(2), 187–205.
32. Noughabi, H. A., & Arghami, N. R. (2011). Testing exponentiality based on characterizations of the exponential distribution. *Journal of Statistical Computation and Simulation, 1*(1), 1–11.
33. Szynal, D. (2010). Goodness-of-fit tests derived from characterizations of continuous distributions. In *Stability in probability* (Vol. 90, pp. 203–223). Warsaw, Poland: Banach Center Publications, Institute of Mathematics, Polish Academy of Sciences.
34. Szynal, D., & Wolynski, W. (2012). Goodness-of-fit tests for exponentiality and Rayleigh distribution. *International Journal of Pure and Applied Mathematics, 78*(5), 751–772.
35. Spiegel, M. R. (1965). *Laplace transforms.* New York, NY: McGraw Hill.
36. Oberhettinger, F., & Badii, L. (1973). *Tables of Laplace transforms.* New York, NY: Springer-Verlag.

3

Reliability, Safety, and Usability Basics

3.1 Introduction

The history of the reliability discipline may be traced back to the early years of the 1930s, when probability concepts were applied to problems associated with electric power systems [1, 2]. However, its real beginning is generally regarded as World War II, when German scientists applied basic reliability concepts to improve the reliability of their V1 and V2 rockets [3]. Today, the reliability field has become a well-developed discipline and has branched out into areas such as software reliability, mechanical reliability, human reliability, and power system reliability [3–5].

The history of the modern safety field goes back to 1868, when a patent was awarded for the first barrier safeguard in the United States [6]. In 1893, the U.S. Congress passed the Railway Safety Act. Nowadays, the safety field has developed into areas such as system safety, workplace safety, medical equipment safety, and software safety [7].

The emergence of usability engineering is deeply embedded in the human factors discipline. The importance of human factors/usability in the design of engineering products/systems can be traced back to 1901 in the contract document of the Army Signal Corps for the development of the Wright Brothers' airplane. The document clearly states that the aircraft be "simple to operate and maintain" [8]. The term *usability engineering* was coined in the mid-1980s, and today there are a large number of publications concerned with usability engineering in the form of books, journal articles, conference proceedings articles, and technical reports [9]. This chapter presents fundamental aspects of reliability, safety, and usability that will be useful in understanding the subsequent chapters in this book.

3.2 Bathtub Hazard Rate Curve

The bathtub hazard rate curve shown in Figure 3.1 is used to describe failure rate for many engineering products/systems/parts, and its name stems

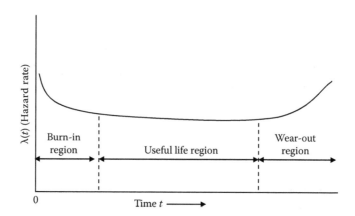

FIGURE 3.1
Bathtub hazard rate curve.

from its shape, which resembles a bathtub. The curve shown in the figure is divided into three regions: burn-in region, useful life region, and wear-out region. These three regions represent three phases that a newly manufactured engineering system or product goes through during its life span.

During the burn-in period, the product hazard rate decreases. Some of the reasons for failures within this region are poor manufacturing methods, poor workmanship, poor quality control, poor processes, substandard materials, and human error [10]. Other terms used for this very region are *debugging region, infant mortality region,* and *break-in region*. During the useful life period, the product hazard rate remains constant, and some of the reasons for the occurrence of failures in this region are as follows [10]:

- Higher random stress than expected
- Abuse
- Undetectable defects
- Natural failures
- Low safety factors
- Human error

Finally, during the wear-out period, the product hazard rate increases, and some of the reasons for the occurrence of failures in this region are as follows [10]:

- Poor maintenance
- Wear due to aging
- Wear due to friction
- Corrosion and creep

- Wrong overhaul practices
- Short designed-in life of the product/item under consideration

The following hazard rate function can be used to represent the bathtub hazard rate curve shown in Figure 3.1 [11]:

$$\lambda(t) = b\alpha(\alpha t)^{b-1} e^{(\alpha t)^b} \tag{3.1}$$

where
$\lambda(t)$ = hazard rate
b = shape parameter
t = time
α = scale parameter

At $b = 0.5$, Equation (3.1) gives the shape of the bathtub hazard rate curve shown in Figure 3.1.

3.3 General Reliability-Related Formulas

A number of general formulas are often used to perform reliability-related analysis. This section presents four such formulas that are based on the reliability function.

3.3.1 Failure (or Probability) Density Function

This is expressed by

$$f(t) = -\frac{dR(t)}{dt} \tag{3.2}$$

where
$R(t)$ = item/product/system reliability at time t
$f(t)$ = failure (or probability) density function

3.3.2 Hazard Rate (or Time-Dependent Failure Rate) Function

This is defined by

$$\lambda(t) = \frac{f(t)}{R(t)} \tag{3.3}$$

where

$\lambda(t)$ = item/product/system hazard rate or time dependent failure rate

By inserting Equation (3.2) into Equation (3.3), we get

$$\lambda(t) = -\frac{1}{R(t)} \cdot \frac{dR(t)}{dt} \tag{3.4}$$

Example 3.1

Assume that the reliability of a computer system is expressed by

$$R(t) = e^{-\lambda t} \tag{3.5}$$

where

$R(t)$ = the computer system reliability at time t
λ = the computer system constant failure rate

Obtain expressions for the computer system failure density function and hazard rate function by using Equations (3.2), (3.4), and (3.5).

By substituting Equation (3.5) into Equation (3.2), we obtain

$$f(t) = -\frac{de^{-\lambda t}}{dt}$$

$$= -(-\lambda e^{-\lambda t})$$

$$= \lambda e^{-\lambda t} \tag{3.6}$$

Inserting Equation (3.5) into Equation (3.4) yields

$$\lambda(t) = -\frac{1}{e^{-\lambda t}} \cdot \frac{de^{-\lambda t}}{dt} = \lambda \tag{3.7}$$

Thus, Equations (3.6) and (3.7) are the expressions for the computer system failure density function and hazard rate function, respectively.

3.3.3 General Reliability Function

This can be obtained with the aid of Equation (3.4). Using Equation (3.4), we write

$$-\lambda(t)dt = \frac{1}{R(t)} dt \tag{3.8}$$

By integrating both sides of Equation (3.8) over the time interval [0,t], we obtain

$$-\int_0^t \lambda(t)dt = \int_1^{R(t)} \frac{1}{R(t)} dR(t) \tag{3.9}$$

since at time $t = 0$, $R(t) = 1$.

Evaluating the right-hand side of Equation (3.9) and then rearranging it yields

$$\ln R(t) = -\int_0^t \lambda(t)\,dt \qquad (3.10)$$

From Equation (3.10), we obtain

$$R(t) = e^{-\int_0^t \lambda(t)\,dt} \qquad (3.11)$$

Equation (3.11) is the general reliability function, which can be used to obtain the reliability of an item whose hazard rate or time-to-failure probability distribution (e.g., exponential, Rayleigh, and Weibull) is known.

Example 3.2

Assume that the hazard rate of a microprocessor is expressed by Equation (3.1). Obtain an expression for the microprocessor's reliability by using Equation (3.11). Substituting Equation (3.1) into Equation (3.11) yields

$$R(t) = e^{-\int_0^t \left\{ b\alpha(\alpha t)^{b-1} e^{(\alpha t)^b} \right\} dt }$$

$$= e^{-\left\{ e^{(\alpha t)^b} - 1 \right\}} \qquad (3.12)$$

Thus, Equation (3.12) is the expression for the microprocessor's reliability.

3.3.4 Mean Time to Failure

Mean time to failure can be obtained by using any of the following three formulas [12]:

$$MTTF = \int_0^\infty R(t)\,dt \qquad (3.13)$$

or

$$MTTF = \lim_{s \to 0} R(s) \qquad (3.14)$$

or

$$MTTF = E(t) = \int_0^\infty tf(t)\,dt \qquad (3.15)$$

where
  MTTF = item's mean time to failure
  t = time

$f(t)$ = failure density function
s = Laplace transform variable
$E(t)$ = expected value
$R(s)$ = Laplace transform of the reliability function, $R(t)$

Example 3.3

Using Equation (3.5), prove that the end results given by Equations (3.13) and (3.14) are identical.
    By substituting Equation (3.5) into Equationd (3.13), we get

$$MTTF = \int_0^\infty e^{-\lambda t}dt$$
$$= \frac{1}{\lambda} \qquad (3.16)$$

By taking the Laplace transform of Equation (3.5), we obtain

$$R(s) = \frac{1}{s+\lambda} \qquad (3.17)$$

Using Equation (3.17) in Equation (3.14) yields

$$MTTF = \lim_{s\to 0} \frac{1}{(s+\lambda)}$$
$$= \frac{1}{\lambda} \qquad (3.18)$$

As Equations (3.16) and (3.18) are exactly the same, it proves that Equations (3.13) and (3.14) give the identical end results.

3.4 Reliability Configurations

In performing reliability analysis, the analyst can encounter computer systems in a variety of different configurations. This section examines some common configurations.

3.4.1 Series Network

This is the simplest reliability configuration or network, and its block diagram is shown in Figure 3.2. Each block in the figure denotes a unit/part. If any one of the units/parts fails, the series system/network fails. In other

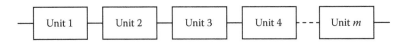

FIGURE 3.2
A series network/system containing *m* units.

words, for the successful operation of the series system, all the system units must work normally.

If we let E_j denote the event that the *j*th unit is successful, then the reliability of the series system shown in Figure 3.2 is given by

$$R_s = P(E_1 E_2 E_3 E_4 \ldots E_m) \tag{3.19}$$

where
  R_s = series system or network reliability
  m = number of units
  $P(E_1 E_2 E_3 E_4 \ldots E_m)$ = probability of occurrence of events E_1, E_2, E_3, E_4, ..., E_m

For independent units, Equation (3.19) becomes

$$R_s = P(E_1)P(E_2)P(E_3)P(E_4) \ldots P(E_m) \tag{3.20}$$

where
  $P(E_j)$ = occurrence probability of event E_j for $j = 1,2,3,4,\ldots,m$

If we let $R_j = P(E_j)$ for $j = 1,2,3,4,\ldots,m$, then Equation (3.20) becomes

$$R_s = R_1 R_2 R_3 R_4 \ldots R_m$$
$$= \prod_{j=1}^{m} R_j \tag{3.21}$$

where
  R_j = reliability of unit *j* for $j = 1,2,3,4,\ldots,m$

Note that because the value of R_j is generally between 0 and 1, the reliability of the series system decreases with the increasing value of *m*.

For constant failure rate, λ_j, of unit *j*, using Equation (3.11), we obtain the following expression for the reliability of the unit *j*:

$$R_j(t) = e^{-\lambda_j t} \tag{3.22}$$

where
  $R_j(t)$ = unit *j* reliability at time *t*

Thus, by substituting Equation (3.22) into Equation (3.21) we get

$$R_s(t) = e^{-\sum_{j=1}^{m} \lambda_j t} \tag{3.23}$$

where
$R_s(t)$ = series system/network reliability at time t

Inserting Equation (3.23) into Equation (3.13) yields the following expression for the series system mean time to failure:

$$\text{MTTF}_s = \int_0^\infty e^{-\sum_{j=1}^{m} \lambda_j t} \, dt$$
$$= \frac{1}{\sum_{j=1}^{m} \lambda_j} \tag{3.24}$$

where
MTTF_s = series system mean time to failure

By substituting Equation (3.23) into Equation (3.4), we obtain the following expression for the series system hazard rate:

$$\lambda_s(t) = -\frac{1}{e^{-\sum_{j=1}^{m} \lambda_j t}} \left[-\sum_{j=1}^{m} \lambda_j \right] e^{-\sum_{j=1}^{m} \lambda_j t}$$
$$= \sum_{j=1}^{m} \lambda_j \tag{3.25}$$

where
$\lambda_s(t)$ = series system hazard rate

Note that the right-hand side of Equation (3.25) is independent of time t. Thus, the left-hand side of Equation (3.25) is simply λ_s, the constant failure rate of the series system. This means that in order to get the series system hazard or constant failure rate, one simply adds up the constant failure rates of each of its units.

Example 3.4

Assume that a computer system is composed of five independent and identical subsystems, and if any one of the subsystems fails, the computer system fails. The failure rate of a subsystem is 0.0008 failures per hour.

Calculate the computer system reliability for a 100-hour mission, mean time to failure, and failure rate.

By substituting the given data values into Equation (3.23), we get

$$R_s(100) = e^{-(0.0008 + 0.0008 + 0.0008 + 0.0008)(100)}$$

$$= 0.7261$$

Using the given data values in Equation (3.24) yields

$$\text{MTTF}_s = \frac{1}{(0.0008 + 0.0008 + 0.0008 + 0.0008)}$$

$$= 312.5 \text{ hours}$$

Finally, by inserting the given data values into Equation (3.25), we obtain

$$\lambda_s = (0.0008 + 0.0008 + 0.0008 + 0.0008)$$

$$= 0.0032 \text{ failures per hour}$$

Thus, the computer system reliability, mean time to failure, and failure rate are 0.7261, 312.5 hours, and 0.0032 failures per hour, respectively.

3.4.2 Parallel Network

This is a quite common network and represents a system with m units operating simultaneously. At least one of these m units must operate normally for the successful operation of the system. The block diagram of an m-unit parallel system is shown in Figure 3.3. Each block in the diagram represents a unit.

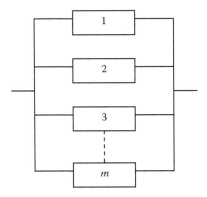

FIGURE 3.3
Block diagram of a parallel system/network with m units.

If we let $\bar{x}_j$ denote the event that the jth unit is unsuccessful, then the parallel system's failure probability is given by

$$F_p = P\left(\bar{x}_1\,\bar{x}_2\,\bar{x}_3\ldots\bar{x}_m\right) \qquad (3.26)$$

where
F_p = failure probability of the parallel system
$P\left(\bar{x}_1\,\bar{x}_2\,\bar{x}_3\ldots\bar{x}_m\right)$ = probability of occurrence of failure events $\bar{x}_1,\bar{x}_2,\bar{x}_3,\ldots,\bar{x}_m$

For independently failing units, Equation (3.26) is written as

$$F_p = P(\bar{x}_1)P(\bar{x}_2)P(\bar{x}_3)\ldots P(\bar{x}_m) \qquad (3.27)$$

where
$P(\bar{x}_j)$ = probability of occurrence of failure event $\bar{x}_j$ for j = 1,2,3,...,m

If we let $F_j = P(\bar{x}_j)$ for j = 1,2,3,...,m then Equation (3.27) becomes

$$F_p = \prod_{j=1}^{m} F_j \qquad (3.28)$$

where
F_j = failure probability of unit j for j = 1,2,3,...,m

Subtracting Equation (3.28) from unit yields the following expression for parallel system/network reliability:

$$R_p = 1 - F_p$$
$$= 1 - \prod_{j=1}^{m} F_j \qquad (3.29)$$

where
R_p = parallel system/network reliability

For constant failure rates of units, subtracting Equation (3.22) from unity and then inserting it into Equation (3.29) yields

$$R_p(t) = 1 - \prod_{j=1}^{m} (1 - e^{-\lambda_j t}) \qquad (3.30)$$

where
$R_p(t)$ = parallel system/network reliability at time t

For identical units, Equation (3.30) becomes

$$R_p(t) = 1 - (1 - e^{-\lambda t})^m \qquad (3.31)$$

where
  λ = unit constant failure rate

Substituting Equation (3.31) into Equation (3.13) yields the following expression for the parallel system/network mean time to failure:

$$MTTF_p = \int_0^\infty \left[1 - \left(1 - e^{-\lambda t}\right)^m \right] dt$$

$$= \frac{1}{\lambda} \sum_{j=1}^{m} \frac{1}{j} \tag{3.32}$$

where
  $MTTF_p$ = parallel system/network mean time to failure

Example 3.5

Assume that a computer system has two independent and identical central processing units in parallel. The failure rate of a central processing unit is 0.0005 failures per hour. Calculate the computer system reliability for a 200-hour mission and mean time to failure.
  By substituting the specified data values into Equation (3.31), we get

$$R_p(200) = 1 - [1 - e^{-(0.0005)(200)}]^2$$

$$= 0.9909$$

Inserting the given data values into Equation (3.32) yields

$$MTTF_p = \frac{1}{(0.005)} \left(1 + \frac{1}{2} \right)$$

$$= 3,000 \text{ hours}$$

Thus, the computer system reliability and mean time to failure are 0.9909 and 3,000 hours, respectively.

3.4.3 *k-out-of-m* Network

In this case, the network or system has m active units, and at least k units out of m active units must work normally for successful operation of the system. The series and parallel networks are special cases of this network, with $k = m$ and $k = 1$, respectively.

For independent and identical units, using the binomial distribution we write the following expression for the reliability of the k-out-of-m network [3]:

$$R_{k/m} = \sum_{j=k}^{m} \binom{m}{j} R^j (1-R)^{m-j} \tag{3.33}$$

where

$$\binom{m}{j} = \frac{m!}{(m-j)!\,j!} \tag{3.34}$$

$R_{k/m}$ = k-out-of-m network/system reliability
R = unit reliability

For constant failure rates of units, using Equation (3.11), Equation (3.33) becomes

$$R_{k/m}(t) = \sum_{j=k}^{m} \binom{m}{j} e^{-j\lambda t} \left(1 - e^{-\lambda t}\right)^{m-j} \tag{3.35}$$

Substituting Equation (3.35) into Equation (3.13) yields

$$\mathrm{MTTF}_{k/m} = \int_0^{\infty} \left[\sum_{j=k}^{m} \binom{m}{j} e^{-j\lambda t} \left(1 - e^{-\lambda t}\right)^{m-j} \right] dt$$

$$= \frac{1}{\lambda} \sum_{j=k}^{m} \frac{1}{j} \tag{3.36}$$

where
  $\mathrm{MTTF}_{k/m}$ = k-out-of-m system/network mean time to failure
  λ = unit constant failure rate

Example 3.6

Assume that a computer system is composed of four identical and independent units in parallel. At least three units must operate normally for the successful operation of the computer system. Calculate the computer system mean time to failure if the unit constant failure rate is 0.0001 failures per hour.

By inserting the specified data values into Equation (3.36), we obtain

$$\mathrm{MTTF}_{3/4} = \frac{1}{(0.0001)} \left[\frac{1}{3} + \frac{1}{4} \right]$$

$$= 5{,}833.33 \text{ hours}$$

Thus, the computer system mean time to failure is 5,833.33 hours.

3.4.4 Standby Redundancy

This is another type of redundancy used to improve system reliability. In this case, the system contains $(n + 1)$ units, as shown in Figure 3.4. Each block in the figure denotes a unit. As shown in the figure, one unit operates, and the n units remain on standby. As soon as the operating unit fails, the switching mechanism detects the failure and turns on one of the standby units. The system fails when all its units fail.

For identical and independent units, perfect switching and standby units, and time-dependent unit failure rate, the reliability of the standby system is expressed by [3]

$$R_{ss}(t) = \sum_{j=0}^{n} \left[\left[\int_0^t \lambda(t)dt \right]^j e^{-\int_0^t \lambda(t)dt} \right] \Big/ j! \tag{3.37}$$

where
$R_{ss}(t)$ = standby system reliability at time t
n = number of standby units
$\lambda(t)$ = unit hazard rate or time-dependent failure rate

For constant unit failure rate (i.e., $\lambda(t) = \lambda$), Equation (3.37) becomes

$$R_{ss}(t) = \sum_{j=0}^{n} (\lambda t)^j e^{-\lambda t} / j! \tag{3.38}$$

By inserting Equation (3.38) into Equation (3.13), we obtain

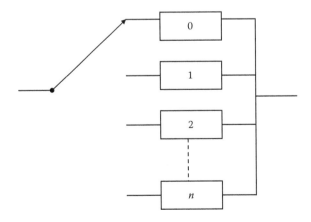

FIGURE 3.4
Block diagram of a standby system with $(n + 1)$ units.

$$\text{MTTF}_{ss} = \int_{0}^{\infty}\left[\sum_{j=0}^{n}(\lambda t)^{j}\,e^{-\lambda t}/j!\right]dt$$

$$= \frac{(n+1)}{\lambda}$$

(3.39)

where

    MTTF_{ss} = standby system mean time to failure

Example 3.7

Assume that a computer system has a standby system containing three identical and independent units: one operating, the other two on standby. The constant failure rate of a unit is 0.0004 failures per hour. Calculate the standby system mean time to failure if the switching mechanism never fails and the standby units remain as good as new in their standby mode.

    By substituting the given data values into Equation (3.39), we get

$$\text{MTTF}_{ss} = \frac{(2+1)}{(0.0004)}$$

$$= 7{,}500 \text{ hours}$$

Thus, the standby system mean time to failure is 7,500 hours.

3.4.5 Bridge Network

This is another configuration that may occur in computer systems. The block diagram of the configuration is shown in Figure 3.5. Each block in the diagram represents a unit.

    For independent units, the Figure 3.5 bridge network reliability is expressed by [13]

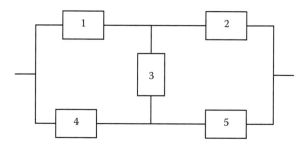

FIGURE 3.5
A five-unit bridge network.

$$R_b = R_1R_3R_5 + R_2R_5 + R_1R_4 + \prod_{j=2}^{4} R_j + 2\prod_{j=1}^{5} R_j - R_1R_2R_4R_5$$

$$- \prod_{j=2}^{5} R_j - \prod_{j=1}^{4} R_j - R_5\prod_{j=1}^{3} R_j - R_1\prod_{j=3}^{5} R_j$$

(3.40)

where
  R_j = reliability of unit j for j = 1,2,3,4,5
  R_b = bridge network/system reliability

For identical units, Equation (3.40) becomes

$$R_b = 2R^5 - 5R^4 + 2R^3 + 2R^2 \tag{3.41}$$

where
  R = unit reliability

For constant failure rates of all five units, using Equations (3.11) and (3.41), we get

$$R_b(t) = 2e^{-5\lambda t} - 5e^{-4\lambda t} + 2e^{-3\lambda t} + 2e^{-2\lambda t} \tag{3.42}$$

where
  $R_b(t)$ = bridge network/system reliability at time t
  λ = unit constant failure rate

By substituting Equation (3.42) into Equation (3.13), we get

$$\text{MTTF}_b = \int_0^{\infty} \left(2e^{-5\lambda t} - 5e^{-4\lambda t} + 2e^{-3\lambda t} + 2e^{-2\lambda t}\right) dt$$

$$= \frac{49}{60\lambda}$$

(3.43)

where
  MTTF_b = bridge network/system mean time to failure

Example 3.8

Assume that five identical and independent units form a bridge network in a computer system. The constant failure rate of a unit is 0.0008 failures per hour. Calculate the bridge network reliability for a 100-hour mission and the mean time to failure.
  By substituting the specified data values into Equation (3.42), we get

$$R_b(100) = 2e^{-5(0.0008)(100)} - 5e^{-4(0.0008)(100)} + 2e^{-3(0.0008)(100)} + 2e^{-2(0.0008)(100)}$$

$$= 1.3406 - 3.6307 + 1.5732 + 1.7042$$

$$= 0.9874$$

Similarly, using the given data value into Equation (3.43) yields

$$MTTF_b = \frac{49}{60(0.0008)}$$

$$= 1,020.83 \text{ hours}$$

Thus, the bridge network reliability and mean time to failure are 0.9874 and 1,020.83 hours, respectively.

3.5 Reliability Evaluation Methods

There are many reliability evaluation methods. This section presents three methods that are useful in evaluating the reliability of computer systems.

3.5.1 Failure Modes and Effect Analysis (FMEA)

This is a commonly used method to analyze engineering systems in regard to reliability, and it may simply be described as an approach to conduct analysis of each system failure mode to examine its effect on the entire system [14]. The history of the FMEA goes back to the early 1950s with the development of flight control systems when the Bureau of Aeronautics of the U.S. Navy developed a reliability-related requirement named "failure analysis" [15].

Subsequently, "failure analysis" became known as "failure effect analysis" and then FMEA [16]. In order to ensure the desired reliability of space systems, the National Aeronautics and Space Administration (NASA) extended FMEA to categorize the effect of each potential failure according to its severity and named it "failure mode effects and criticality analysis" (FMECA) [17].

Seven main steps used to perform FMEA are as follows [18]:

Step 1: Define system boundaries and all related requirements in detail.

Step 2: List all system subsystems and parts.

Step 3: List all possible failure modes and identify and describe the part/component under consideration.

Step 4: Assign appropriate probability or failure rate to each part/component failure mode.

Step 5: List effect(s) of each failure mode on subsystems and the plant.

Step 6: Enter appropriate remarks for each failure mode.

Step 7: Review each critical failure mode and take necessary measures.

Additional information on FMEA is available in the literature [3, 19].

3.5.2 Fault Tree Analysis (FTA)

This is a widely used method in the industrial sector, particularly in the area of nuclear power generation, to perform reliability analysis of engineering systems. The method was developed at the Bell Telephone Laboratories in the early 1960s to analyze the Minuteman launch control system [20].

FTA is concerned with failure or fault events, and a fault tree may simply be described as a logical representation of the primary faults events that lead to the occurrence of a specified undesirable fault event called the "top event." Furthermore, a fault tree is depicted using a tree structure with OR, AND, etc., logic gates.

3.5.2.1 Common Fault Tree Symbols

The FTA method uses many symbols to develop fault trees of engineering systems [18, 21]. The four most commonly used symbols are shown in Figure 3.6. Each of these symbols is described as follows:

Circle: This denotes a basic fault event or the failure of an elementary part or component. The event's probability of occurrence and failure and repair rates are generally obtained from empirical data.

Rectangle: This represents a fault event that results from the combination of fault events through the input of a logic gate.

OR gate: This denotes that an output fault event occurs if any one or more of the input fault events occur.

AND gate: This denotes that an output fault event occurs if and only if all the input fault events occur.

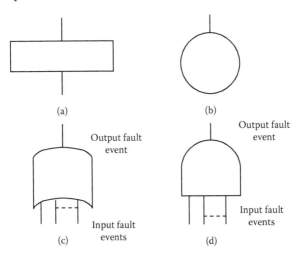

FIGURE 3.6
Four common fault tree symbols: (a) rectangle, (b) circle, (c) OR gate, (d) AND gate.

3.5.2.2 Steps Involved in Developing a Fault Tree

The following five basic steps are normally followed to develop a fault tree [18, 22]:

Step 1: Define the undesired event (i.e., the top event) of the system to be studied.

Step 2: Understand thoroughly the system under consideration and its intended application.

Step 3: Determine the higher-order functional fault events in order to obtain the predefined cause of the system fault condition. Furthermore, continue FTA to determine the appropriate logical interrelationship of lower-level fault events that can cause their occurrence.

Step 4: Construct a fault tree of logical relationships among input fault events.

Step 5: Evaluate the final fault tree qualitatively/quantitatively.

Example 3.9

Assume that a windowless room has three light bulbs and one switch. The room can only be dark if all the three bulbs burn out, the switch fails to close, or there is no electricity. Develop a fault tree, using Figure 3.6 symbols, for the undesired event (i.e., top event) "dark room"—the room without light.

A fault tree for this example is shown in Figure 3.7. Note that each fault event in this figure is labeled as $E_1, E_2, E_3, \ldots, E_9$.

3.5.2.3 Probability Evaluation of Fault Trees

The probability of occurrence of the top event can be calculated when the occurrence probability of basic fault events is known. This requires first calculating the occurrence probability of output fault events of lower and intermediate logic gates such as OR and AND. The probability of occurrence of the output fault events of OR and AND gates is obtained, respectively, as follows [3, 18, 21]:

OR gate: The occurrence probability of the OR gate output fault event Y_0 is expressed by

$$P(Y_0) = 1 - \prod_{i=1}^{m} \left\{ 1 - P(Y_i) \right\} \tag{3.44}$$

where
m = number of independent input fault events
$P(Y_0)$ = probability of occurrence of the OR gate output fault event, Y_0
$P(Y_i)$ = probability of occurrence of input fault event Y_i for $i = 1, 2, 3, \ldots, m$

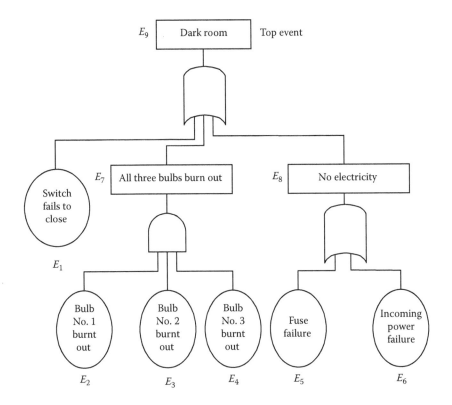

FIGURE 3.7
Fault tree for the occurrence of undesired event (top event): dark room.

AND gate: The occurrence probability of the AND gate output fault event X_0 is expressed by

$$P(X_0) = \prod_{i=1}^{m} P(X_i) \tag{3.45}$$

where
  m = number of independent input fault events
  $P(X_0)$ = probability of occurrence of the AND gate output fault event, X_0
  $P(X_i)$ = probability of occurrence of the input fault event X_i for $i = 1, 2, 3, \ldots, m$

Example 3.10

Assume that in Figure 3.7, the occurrence probability of basic fault events $E_1, E_2, E_3, E_4, E_5,$ and E_6 are 0.07, 0.05, 0.05, 0.05, 0.04, and 0.06, respectively. Calculate the probability of occurrence of the top event (dark room, E_9) and then show the fault tree in Figure 3.7 with calculated and given fault-occurrence probability values.

By substituting the given data values into Equation (3.44), the probability of occurrence of event E_8 (i.e., no electricity) is

$$P(E_8) = 1 - [1 - P(E_5)][-P(E_6)]$$

$$= 1 - [1 - 0.04][1 - 0.06]$$

$$= 0.0976$$

Similarly, using the given data values in Equation (3.45), the probability of occurrence of event E_7 (i.e., all three bulbs burnt out) is

$$P(E_7) = P(E_2)P(E_3)P(E_4)$$

$$= (0.05)(0.05)(0.05)$$

$$= 0.000125$$

Using the above calculated values and the given data value in Equation (3.44) yields

$$P(E_9) = 1 - [1 - P(E_1)][1 - P(E_7)][1 - P(E_8)]$$

$$= 1 - [1 - 0.07][1 - 0.000125][1 - 0.0976]$$

$$= 0.1608$$

where
$P(E_9)$ = occurrence probability of fault event E_9 (i.e., dark room)

Thus, the probability of having no light in the room is 0.1608. The Figure 3.7 fault tree with calculated and given fault-event occurrence probability values is shown in Figure 3.8.

3.5.3 Markov Method

This is a widely used method in the industrial sector to perform various types of reliability analysis and is named after a Russian mathematician, Andrei Andreyevich Markov (1856–1922). The method is quite useful to model systems with constant failure and repair rates and is based on the following three assumptions [23]:

- The probability of transition from one system state to another in the finite time interval Δt is given by $\theta \Delta t$, where θ is the transition rate (e.g., constant failure or repair rate of an item) from one system state to another.

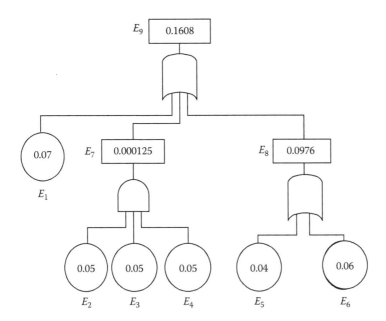

FIGURE 3.8
Fault tree of Figure 3.7 with calculated and given fault-event occurrence probability values.

- The occurrences are independent of each other.
- The probability of occurrence of more than one transition in time interval Δt from one state to another is very small or negligible (e.g., $(\theta \Delta t)(\theta \Delta t) \to 0$).

The application of this method is demonstrated through the following example:

Example 3.11

A computer system can either be in a working state or a failed state. Its constant failure and repair rates are λ_c and μ_c, respectively. The computer system state space diagram is shown in Figure 3.9. The numerals in the boxes denote the computer system state. Using the Markov method, develop expressions for computer system time-dependent and steady-state availabilities and unavailabilities.

With the aid of the Markov method, we write down the following equations for the diagram in Figure 3.9:

$$P_0(t + \Delta t) = P_0(t)(1 - \lambda_c \Delta t) + P_1(t)\mu_c \Delta t \qquad (3.46)$$

$$P_1(t + \Delta t) = P_1(t)(1 - \mu_c \Delta t) + P_1(t)\lambda_c \Delta t \qquad (3.47)$$

where
t = time

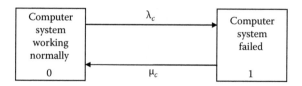

FIGURE 3.9
Computer system state space diagram.

$\lambda_c \Delta t$ = probability of computer system failure in finite time interval Δt
$\mu_c \Delta t$ = probability of computer system repair in finite time interval Δt
$(1 - \lambda_c \Delta t)$ = probability of no failure in finite time interval Δt
$(1 - \mu_c \Delta t)$ = probability of no repair in finite time interval Δt
$P_0(t + \Delta t)$ = probability of the computer system being in working state 0 at time $(t + \Delta t)$
$P_1(t + \Delta t)$ = probability of the computer system being in failed state 1 at time $(t + \Delta t)$
$P_j(t)$ = probability that the computer system is in state j at time t for $j = 0, 1$

In the limiting case, Equations (3.46) and (3.47) become

$$\frac{dP_0(t)}{dt} + \lambda_c P_0(t) = P_1(t)\mu_c \tag{3.48}$$

$$\frac{dP_1(t)}{dt} + \mu_c P_1(t) = P_0(t)\lambda_c \tag{3.49}$$

At time $t = 0$, $P_0(0) = 1$ and $P_1(0) = 0$
Solving Equations (3.48) and (3.49), we obtain

$$A_c(t) = P_0(t) = \frac{\mu_c}{(\lambda_c + \mu_c)} + \frac{\lambda_c}{(\lambda_c + \mu_c)} e^{-(\lambda_c + \mu_c)t} \tag{3.50}$$

$$U_c(t) = P_1(t) = \frac{\lambda_c}{(\lambda_c + \mu_c)} - \frac{\lambda_c}{(\lambda_c + \mu_c)} e^{-(\lambda_c + \mu_c)t} \tag{3.51}$$

where
$A_c(t)$ = computer system time-dependent availability
$U_c(t)$ = computer system time-dependent unavailability

By letting time t go to infinity (∞) in Equations (3.50) and (3.51), we obtain

$$A_c = \lim_{t \to \infty} A_c(t) = \frac{\mu_c}{(\lambda_c + \mu_c)} \tag{3.52}$$

$$U_c = \lim_{t \to \infty} U_c(t) = \frac{\lambda_c}{(\lambda_c + \mu_c)} \tag{3.53}$$

where
A_c = computer system steady-state availability
U_c = computer system steady-state unavailability

Thus, the computer system time-dependent and steady-state availabilities are given by Equations (3.50) and (3.52), respectively. Similarly, the computer system time-dependent and steady-state unavailabilities are given by Equations (3.51) and (3.53), respectively.

Example 3.12

Assume that the constant failure and repair rates of a computer system are 0.0004 failures per hour and 0.08 repairs per hour, respectively. Calculate the computer system availability for a 100-hour mission.

By substituting the given data values into Equation (3.50), we get

$$A_c(100) = \frac{0.08}{(0.0004 + 0.08)} + \frac{0.0004}{(0.0004 + 0.08)} e^{-(0.0004 + 0.08)(100)}$$

$$= 0.9950$$

Thus, the computer system availability is 0.9950 or 99.5%.

3.6 Need for Safety and the Role of Engineers in Designing for Safety

The desire for safety and security is basic to the human condition. Early humans took appropriate measures to guard against naturally caused hazards, and as early as 2000 BCE, a Babylonian ruler named Hammurabi developed a set of laws—the Code of Hammurabi—to mediate the cost of damages resulting from human actions. This code included clauses on various items, including monetary damages against individuals who caused injury to others and allowable fees for involved physicians [6, 22, 24].

In today's world, safety remains an important issue, and the focus is on preventable deaths and injuries. For example, in 1996 the U.S. National Safety Council (NSC) reported 93,400 deaths and a large number of disabling injuries due to accidents in the United States [25]. The estimated cost of these accidents was around $121 billion. There is a clear need and a demand for better safety, and efforts to improve safety are driven by government regulations, an increasing number of lawsuits, and public pressure.

In modern times, the problems concerned with the safety of engineering systems/products may be traced back to the railroads. For example, the day Stephenson's first railroad was dedicated, a railroad accident killed a prominent English legislator [26]. The following year, one person was killed and a number of fuel servers were badly injured when the boiler of the first locomotive built in the United States exploded [26, 27].

Today, engineering systems/products have become highly complex and sophisticated, and their safety has become a challenging issue. Due to competition and other factors, engineers are pressured to complete new designs rapidly and at lower costs. In turn, this usually leads to more design deficiencies, errors, and causes of accidents. As per Hammer and Price [26], a design-related deficiency may occur because a design:

- Is confusing, unfinished, or wrong
- Violates the usual tendencies/capabilities of potential users
- Incorporates poor warning mechanisms rather than providing a safe design to eradicate hazards
- Places an unreasonable level of stress on potential operators
- Creates an unsafe characteristic
- Relies on the user of a product/item to avoid an accident
- Creates an arrangement of operating controls and other devices that considerably increases reaction time in emergency conditions or is conducive to a variety of errors
- Does not properly consider or determine the error, action, failure, or omission consequences
- Overlooks the need to warn properly of a potential hazard
- Overlooks the need to eliminate or reduce human errors
- Overlooks the need to prescribe an appropriate operational procedure in situations where a hazard may exist
- Overlooks the need to foresee an unexpected use of an item or its possible consequences
- Overlooks the need to provide an appropriate level of protection in the personal protective equipment of a worker/user

3.7 Classifications of Product Hazards

There are many types of product hazards. These may be categorized under the following six classifications [22, 28]:

Classification I: Energy hazards. These hazards may be grouped under two categories: potential energy and kinetic energy. The potential energy hazards are associated with parts that store energy. Some examples of these parts are springs, counterbalancing weights, compressed-gas receivers, and electronic capacitors. During the equipment servicing process, such hazards are very important because stored energy can result in serious injuries when released suddenly.

The kinetic energy hazards are associated with parts that have energy because of their motion. Such parts include flywheels, fan blades, and loom shuttles. Any item that interferes with the motion of such parts can experience significant damage.

Classification II: Electrical hazards. These hazards have two main components: shock hazard and electrocution hazard. The major hazard to a product/item stem from electrical faults, often known as short circuits.

Classification III: Human factors hazards. These hazards are concerned with poor design with respect to human factors such as physical strength, intelligence, height, visual angle, weight, computational ability, length of reach, visual acuity, etc.

Classification IV: Kinematic hazards. These hazards are concerned with situations where parts come together while moving and result in possible cutting, crushing, or pinching of any item caught between them.

Classification V: Environmental hazards. These hazards may be grouped under two categories: internal hazards and external hazards. The internal hazards are concerned with the changes in the surrounding environmental conditions that result in an internally damaged product/item. These hazards can be eliminated or minimized by considering factors such as extremes of temperatures, vibrations, electromagnetic radiation, atmospheric contaminants, illumination level, and ambient noise levels during the design phase.

The external hazards are the very hazards posed by the product/item in question during its life span. Some examples of these hazards are disposal hazards, maintenance hazards, and service-life operation hazards.

Classification VI: Misuse-and-abuse hazards. These hazards are associated with the usage of a product/item by humans. Product/item misuse can lead to serious injuries, and abuse can also lead to injuries or hazardous situations. Some of the causes for the abuse are lack of proper maintenance and poor operating practices.

3.8 Human Factors Basics for Engineering Usability

Probably the main reason for the existence of the human factors discipline is that humans keep making errors while using machines. Otherwise, it would be quite difficult to justify its existence. Nonetheless, this section presents some important aspects of human factors basics that are useful for engineering usability.

3.8.1 Comparisons of Humans' and Machines' Capabilities/Limitations

During the design of engineering systems, decisions may have to be made about assigning certain functions to humans or machines. Under such circumstances, a good knowledge of the capabilities/limitations of humans and machines is essential; otherwise, the right decisions may not be made. Some important comparisons of humans' and machines' capabilities and limitations are presented in Table 3.1 [9, 29].

TABLE 3.1

Comparisons of Capabilities/Limitations of Humans and Machines

No.	Humans	Machines
1	Are very flexible with respect to task performance	Are relatively quite inflexible
2	Are very capable of making inductive decisions under novel conditions	Have very small or no induction capabilities at all
3	Are subjected to social environments of all kinds	Are free of social environments of all kinds
4	Are subjected to factors such as motion sickness, Coriolis effect, and disorientation	Are completely free of such effects
5	Performance-related efficiency is affected by anxiety	Are independent of this shortcoming
6	Are limited to a certain degree in channel capacity	Can have unlimited amount of channel capacities
7	Have tolerance for factors such as ambiguity, uncertainty, and vagueness	Are limited in tolerance in regard to factors such as these
8	Have restricted short-term memory for factual matters	Can have unlimited amount of short-term memory, but its affordability is a limiting factor
9	Are subjected to stress because of interpersonal or other problems	Are completely free of such problems
10	Have excellent memory	Are very costly to have the same capability
11	Are unsuitable to perform tasks such as data coding, transformation, or amplification	Are extremely useful to perform tasks such as these
12	Are a poor monitor of events that do not occur often	Possess an option to be designed to reliably detect rarely occurring events
13	Are often subjected to depart from following an optimum strategy	Always follow the designed-in strategy
14	Are subjected to deterioration in performance because of fatigue and boredom	Are not affected by factors such as these, but their performance is subjected to deterioration due to lack of calibration or wear
15	Have relatively easy maintenance needs	Maintenance-related problems become quite serious with the increase in complexity

3.8.2 Typical Human Behaviors and Their Corresponding Design Considerations

Over the years, researchers working in this area have identified many typical human behaviors. Some of the typical or expected human behaviors—and their corresponding design considerations in parentheses—are as follows [9, 29]:

- Humans generally consider manufactured products as being safe (place emphasis on designing products in such a way that they cannot be used wrongly).
- Humans have become accustomed to the meanings of certain colors (strictly observe existing color-coding standards during the design process).
- Humans expect that to turn on the power, the electrically powered switches have to move to the upward direction, or to the right direction, etc. (design such items as per human expectations).
- Humans get easily confused with unfamiliar items (avoid designing totally unfamiliar items).
- Humans will often use their sense of touch to explore or test the unknown (pay close attention to this factor during the design process, particularly to the handling aspect of the product under consideration).
- Humans expect that valve handles/faucets will rotate counterclockwise to increase the flow of gas, liquid, or steam (design items such as these according to the expectations of humans).
- Humans generally have very little knowledge about their physical shortcomings (develop proper design by carefully considering relevant human shortcomings and basic characteristics).
- Humans frequently tend to hurry (develop appropriate design so that it effectively takes the human hurry element into consideration).
- The attention of humans is drawn to items such as bright lights, bright and vivid colors, flashing lights, and loud noise (design in stimuli of proper intensity when attention needs stimulation).

3.8.3 Human Sensory Capacities

Humans possess five useful sensors: smell, hearing, sight, taste, and touch, as shown in Figure 3.10 [30]. A better understanding of their sensory capacities can be very useful in reducing the occurrence of human errors in engineering systems. The effects of touch, sight, vibration, and noise are discussed in the following subsections [30].

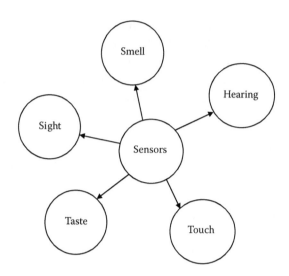

FIGURE 3.10
Useful sensors possessed by humans.

3.8.3.1 Touch

The sense of touch is quite closely associated with the human ability to interpret visual and auditory stimuli. The sensory cues received by the muscles and skin can be used to a certain degree for sending messages to the brain, thus relieving the ears and the eyes of a significant part of the workload. In circumstances when human users are expected to rely completely on their touch sensors, different types of knob shapes could be adopted for use.

The application of the touch sensor in technical tasks is not new; in fact, it has been used for hundreds of years by craft workers to detect surface-related roughness and irregularities in their work. Over the years, many studies have clearly indicated that the detection accuracy of surface irregularities increases dramatically when one moves an intermediate, a piece of paper or thin cloth, over the object surface instead of just using bare fingers [31].

3.8.3.2 Sight

Sight is stimulated by the electromagnetic radiation of certain wavelengths, often called the visible portion of the electromagnetic spectrum. The spectrum's various areas, as seen by the human eye, appear to vary in brightness. For example, in the daylight, the eyes of the human are most sensitive to greenish-yellow light with a wavelength of about 5,500 angstroms [30]. Furthermore, the eyes see differently from different angles. They perceive all colors when they are looking straight ahead, but as the viewing angle increases, the color perception begins to decrease.

Some of the other factors directly or indirectly concerned with color are as follows [30]:

- Staring at a certain colored light and then glancing away may result in the reversal of color in one's brain. For example, after staring at a red or green light and then glancing away, the signal to one's brain may completely reverse the color.
- Color-weak individuals do not see colors in a similar way as normal people do.
- At night or in poorly illuminated areas, color differences are very minimal. In particular, for a small point source (e.g., a small warning light) or from a distance, it is quite impossible to make a clear distinction between yellow, green, orange, and blue. In fact, all of these colors will appear to be white.

Three sight-related guidelines considered very important are as follows [30]:

- Select the right color so that color-weak individuals do not get confused.
- Make use of red filters, if possible, with wavelengths greater than 6,500 angstroms.
- Avoid relying too much on color in situations where critical tasks may be carried out by fatigued people.

3.8.3.3 Vibration

The existence of vibration could be quite detrimental to the performance of mental and physical activities by humans. There are many parameters of vibrations, including velocity, acceleration, frequency, and amplitude. In particular, low-frequency and large-amplitude vibrations contribute to problems such as fatigue, headaches, eyestrain, motion sickness, and interference with one's ability to interpret and read instruments [30]. These symptoms become less pronounced as the frequency of vibrations increases and the amplitude decreases. Note that high-frequency and low-amplitude vibrations can also be quite fatiguing.

Some of the important guidelines to decrease the effects of motion and vibration are presented as follows [30, 32]:

- Eliminate altogether, if possible, vibrations greater than 0.8-mil (20.32 micron) amplitude.
- Resist shocks and vibrations through effective design measures or isolate them with the aid of items such as springs, cushioned mountings, shock absorbers, and fluid couplings.
- Use damping materials or cushioned seats to lower vibrations transmitted to a seated person and avoid vibrations of frequencies 3 to 4 Hz, as this is a vertically seated individual's resonant frequency.

3.8.3.4 Noise

Noise may simply be described as sounds that lack coherence, and its effects on people are quite difficult to measure exactly. Reactions of humans to noise extend beyond the auditory systems (i.e., to feelings such as irritability, fatigue, boredom, or well-being). Moreover, excessive noise can lead to problems such as the reduction in the efficiency of workers, loss of hearing if exposed for long periods, and the adverse effects on activities that require a rather high degree of muscular coordination and precision or intense concentration.

Generally, two main physical characteristics (i.e., frequency and intensity) are used in describing noise. Intensity is normally measured in decibels (dB), and a person exposed to greater than 80 dB of noise can suffer from permanent/temporary loss of hearing. The levels of intensity for noise sources such as heavy traffic, normal conversation, household ventilation fan, quiet residential area, voice whisper, and motion picture sound studio are 70, 60, 56, 40, 20, and 10 dB, respectively [8, 33].

In regard to frequency, the ears of humans are most sensitive to frequencies in the range of 600–900 Hz, and they have the capacity to detect sounds of frequencies from 20 to 20,000 Hz. People usually suffer a major loss of hearing when they are exposed to frequencies of noise between 4,000 and 6,000 Hz for a long period of time [8, 30].

Problems

1. Describe the bathtub hazard rate curve.
2. Prove Equation (3.11) by using Equation (3.4).
3. Write down three different formulas to obtain mean time to failure.
4. Assume that a computer system is composed of four identical and independent subsystems and if anyone of the subsystems fails the computer system fails. The failure rate of a subsystem is 0.0002 failures per hour. Calculate the computer system reliability for a 200-hour mission, mean time to failure, and failure rate.
5. Describe failure modes and effect analysis (FMEA).
6. What are the main steps involved in developing a fault tree?
7. Discuss product hazards and their main classifications.
8. Discuss at least eight typical human behaviors.
9. Discuss the effects of noise, sight, and touch.
10. Assume that the constant failure and repair rates of a computer system are 0.0007 failures per hour and 0.04 repairs per hour, respectively. Calculate the computer system unavailability for a 600-hour mission.

References

1. Smith, S. A. (1934). Service reliability measured by probabilities of outage. *Electrical World, 103*, 371–374.
2. Lyman, W. J. (1933). Fundamental consideration in preparing a master system plan. *Electrical World, 101*, 778–792.
3. Dhillon, B. S. (1999). *Design reliability: Fundamentals and applications.* Boca Raton, FL: CRC Press.
4. Dhillon, B. S. (1983). *Power system reliability, safety, and management.* Ann Arbor, MI: Ann Arbor Science Publishers.
5. Dhillon, B. S. (1986). *Human reliability: With human factors.* New York, NY: Pergamon Press.
6. Goetsch, D. L. (1996). *Occupational safety and health.* Englewood Cliffs, NJ: Prentice Hall.
7. Dhillon, B. S. (2003). *Engineering safety: Fundamentals, techniques, and applications.* River Edge, NJ: World Scientific Publishing.
8. Department of Defense. (1976). *Engineering design handbook: Maintainability engineering theory and practice* (AMCP 706–133). Washington, DC: Author.
9. Dhillon, B. S. (2004). *Engineering usability: Fundamentals, applications, human factors, and human error.* Stevenson Ranch, CA: American Scientific Publishers.
10. Kapur, K. C. (1982). Reliability and maintainability. In G. Salvendy (Ed.), *Handbook of industrial engineering* (pp. 8.5.1–8.5.34). New York, NY: John Wiley and Sons.
11. Dhillon, B. S. (1981). Life distributions. *IEEE Transactions on Reliability, 30*(5), 457–460.
12. Dhillon, B. S. (1988). *Mechanical reliability: Theory, models, and applications.* Washington, DC: American Institute of Aeronautics and Astronautics.
13. Lipp, J. P. (1957). Topology of switching elements vs. reliability. *Trans. IRE Reliability Quality Control, 7*, 21–34.
14. Omdahl, T. P. (Ed.). (1988). *Reliability, availability, and maintainability (RAM) dictionary.* Milwaukee, WI: American Society for Quality Control (ASQC) Press.
15. Department of the Navy. (1955). *General specification for design, installation, and test of aircraft flight control systems* (MIL-F-18372 [Aer]) (Para. 3.5.2.3). Washington, DC: Bureau of Naval Weapons.
16. Coutinho, J. S. (1964). Failure effect analysis. *Trans. of the New York Academy of Sciences, 26*(Series II), 564–584.
17. Jordan, W. E. (1972). Failure modes, effects, and criticality analyses. *Proceedings of the Annual Reliability and Maintainability Symposium*, 30–37.
18. Dhillon, B. S., & Singh, C. (1981). *Engineering reliability: New techniques and applications.* New York, NY: John Wiley and Sons.
19. Palady, P. (1995). *Failure modes and effect analysis.* West Palm Beach, FL: PT Publications.
20. Haasl, D. F. (1965, June). Advanced concepts in fault tree analysis. *Proceedings of the System Safety Symposium*, University of Washington, Seattle.
21. U.S. Nuclear Regulatory Commission. (1981). *Fault tree handbook* (Report No. NUREG-0492). Washington, DC: Author.

22. Dhillon, B. S. (2005). *Reliability, quality, and safety for engineers*. Boca Raton, FL: CRC Press.
23. Shooman, M. L. (1968). *Probabilistic reliability: An engineering approach*. New York, NY: McGraw Hill.
24. Ladon, J. (Ed.). (1986). *Introduction to health and safety*. Chicago, IL: National Safety Council (NSC).
25. National Safety Council. (1996). *Accidental facts*. Chicago, IL: Author.
26. Hammer, W., & Price, D. (2001). *Occupational safety management and engineering*. Upper Saddle River, NJ: Prentice Hall.
27. Operator safety. (1974, May). *Engineering, 1974*(May), 358–363.
28. Hunter, T. A. (1992). *Engineering design for safety*. New York, NY: McGraw Hill.
29. Woodson, W. E. (1981). *Human factors design handbook*. New York, NY: McGraw Hill.
30. U.S. Army Material Command. (1972). *Engineering design handbook: Maintainability guide for design* (AMCP 706-134). Alexandria, VA: Author.
31. Lederman, S. (1978). Heightening tactile impression of surface texture. In G. Gordon (Ed.), *Active touch* (pp. 40–45). New York, NY: Pergamon Press.
32. Altman, J. W., Marchese, A. C., & Marchiando, B. W. (1961). *Guide to design of mechanical equipment for maintainability* (Report No. ASD-TR-61–381). Wright-Patterson Air Force Base, OH: Air Force Systems Command.
33. McCormick, E. J., & Sanders, M. S. (1982). *Human factors in engineering and design*. New York, NY: McGraw Hill.

4

Computer System Reliability Basics

4.1 Introduction

Over the years, computer applications have increased at an alarming rate, ranging from personal use to control aerospace and nuclear systems. As computer failures can affect our day-to-day life, their reliability has become very important to the population at large. Furthermore, the reliability of computer systems used in areas such as defense, nuclear power generation, and aerospace is of utmost importance because their failures could be catastrophic and very costly.

The history of computer system reliability may be traced back to the late 1940s and 1950s [1–4]. For example, in 1956 Von Neumann [4] proposed the triple modular redundancy (TMR) scheme to improve system reliability. Needless to say, over the years a large number of publications have appeared in the form of journal articles, conference proceedings articles, technical reports, and books to improve computer system reliability [5, 6]. This chapter presents various important aspects of computer system reliability basics extracted from the published literature.

4.2 Hardware Reliability Versus Software Reliability

As it is very important to have a clear understanding of the differences between computer software and hardware reliability, Table 4.1 presents comparisons of a number of important areas [7–10].

TABLE 4.1

Software and Hardware Reliability Comparisons

No.	Software Reliability	Hardware Reliability
1	Software failure is caused by programming error	A hardware failure is generally due to physical effects
2	Redundancy may not be effective	Normally redundancy is effective
3	Does not wear out	Wears out
4	Preventive maintenance has no meaning in software	Preventive maintenance is carried out to inhibit failures
5	Corrective maintenance is basically redesign	The failed item/system is repaired by carrying out corrective maintenance
6	Mean time to repair (MTTR) has no significance	Mean time to repair (MTTR) has significance
7	Software does not fail as per the bathtub hazard rate curve	Many hardware items fail as per the bathtub hazard rate curve
8	Interfaces are conceptual	Interfaces are visual
9	Software reliability still lacks well-developed theory and mathematical concepts	Hardware reliability has well-developed theory and mathematical concepts
10	Obtaining good failure-related data is a problem	Obtaining good failure-related data is a problem
11	The software reliability field is relatively new	The hardware reliability field is quite well established, particularly with respect to electronics
12	It is impossible to repair software failures by using spare modules	It is quite possible to repair hardware by using spare modules

4.3 Major Sources of Computer Failures and Issues in Computer System Reliability

There are many causes of computer failures. The major sources that lead to computer failures are as follows [9, 11, 12]:

- Peripheral device failures
- Processor and memory failures
- Human errors
- Communication network failures
- Environmental and power failures
- Mysterious failures
- Gradual erosion of the database
- Saturation

The first six of these major sources of computer failures are described as follows:

- Peripheral device failures are important, but they rarely lead to a system shutdown. The commonly occurring errors in peripheral devices are intermittent or transient, and the electromechanical nature of the peripheral devices is the usual reason for their failure.

- Normally, processor errors/failures are catastrophic, but their occurrence is quite rare, as there are times when the central processor malfunctions and fails to execute instructions correctly because of a "dropped bit." Nowadays, the occurrence of memory parity errors is quite rare due to improvements in hardware reliability, and these errors are not necessarily fatal.

- Human errors generally occur due to operator mistakes and oversights. Operator errors frequently take place during shutting down, starting up, and running the computer system.

- Communication network failures are concerned with intermodule communication, and many of these failures are generally of a transient nature. Around two-thirds of errors in communication lines can be detected with the application of "vertical parity" logic.

- Environmental failures occur due to causes such as electromagnetic interference, fires, earthquakes, and air conditioning equipment failure. In the case of power failures, factors such as transient fluctuations in voltage or frequency and total power loss from the local utility company are the causes of their occurrence.

- Mysterious failures are never categorized properly in real-time systems because they occur unexpectedly. For example, when a normally operating system stops functioning at once without indicating any problem (i.e., hardware, software, etc.) at all, the failure is referred to as a mysterious failure.

There are many issues concerned with computer system reliability. Some of the important factors to consider are as follows [6, 13]:

- Failures in the area of computer systems are highly varied in character. For example, a part or component used in a computer system may experience a transient fault due to its environment, or it may fail permanently.

- Prior to the installation and production phases, it could be quite difficult to detect errors associated with hardware design at the lowest system levels. It is quite possible that mistakes in hardware design may result in situations where operational errors due to such oversights are impossible to distinguish from the ones due to transient physical faults.

- The main parts/components of computers are the logic elements, which have quite troublesome reliability-related features. In many situations it is impossible to determine properly the reliability of such elements, and their defects cannot be healed properly.
- Generally, the most powerful type of self-repair in computers is dynamic fault tolerance, but it is very difficult to analyze. Nonetheless, for certain applications it is very important and cannot be ignored.
- Modern computers consist of redundancy schemes for fault tolerance, and advances over the years have brought various types of improvements, but there are still many theoretical and practical difficulties that remain to be overcome.

4.4 Fault Classifications and Computer Reliability Measures

Generally, for the purpose of computer system reliability modeling and evaluation, an effective approach for classifying computer-related faults is on the basis of their duration. Thus, the faults may be classified under the following two categories [14, 15]:

Transient faults. These types of faults are due to the temporary malfunction of components/parts or by external interference such as electrical noise, power dips, and glitches. These faults are of limited duration, and although they need restoration, they do not involve any replacement or repair. These types of faults are characterized by arrival modes and the duration of transients [14, 15].

Permanent faults. These types of faults are frequently due to catastrophic failures of components/parts. In this case, the failure of the component/part is irreversible and permanent and requires replacement or repair. These faults are characterized by long duration and have a failure rate that depends on the surrounding environment. For example, a part or component will generally have a different failure rate in power-off and power-on conditions [15, 16].

There are many measures used in the area of computer system reliability, but they can be divided into two groups as follows [6]:

Group I. This group contains measures that are suitable for configurations such as standby, hybrid, and massively redundant systems. The measures are mission time, system reliability, mean time to failure, and system availability. Note that to evaluate gracefully degrading systems, these measures may not be sufficient.

Group II. This group contains five new measures to handle gracefully degrading systems. Each of these five measures is defined as follows:

Computation reliability. This is the failure-free probability that the system will, without an error, execute a task of length, say y, started at time t.

Mean computation before failure. This is the expected amount of computation available on the system prior to failure.

Computation threshold. This is the time at which a certain value of computation reliability is reached for a task whose length is, say y.

Computation availability. This is the expected computation capacity of the system at a given time t.

Capacity threshold. This is the time at which a certain value of computation availability is reached.

4.5 Fault Masking

In the area of fault-tolerant computing, the term *fault masking* is used in the sense that a system with redundancy can tolerate a number of malfunctions or failures before its own failure. The implication of the term *masking* is that some kind of problem has surfaced somewhere within the framework of a digital system, but because of design, the problem does not affect the overall operation of the system. Modular redundancy is probably the best-known fault masking method.

4.5.1 Triple Modular Redundancy (TMR)

In this case, three identical redundant modules or units carry out the same task simultaneously, and a voter compares the outputs from all the three modules and sides with the majority. The TMR system fails only when two or more modules/units fail or the voter fails. More clearly, the TMR system can tolerate the failure of a single module or unit. A block diagram of the TMR system with voter is shown in Figure 4.1 [6, 9].

The TMR system or scheme was first proposed in 1956, and an important example of its application is the Saturn V launch vehicle computer [4, 9]. The vehicle computer used TMR with voters in the central processor and duplication in the main memory [9, 17].

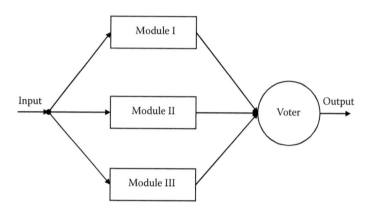

FIGURE 4.1
Block diagram for TMR system with voter.

For independent units, the reliability of the TMR system with voter (shown in Figure 4.1) is given by the following equation [9]:

$$R_{TV} = (3R^2 - 2R^3)R_v \qquad (4.1)$$

where
R_{TV} = TMR system with voter reliability
R = module/unit reliability
R_v = voter reliability

With a perfect voter (i.e., $R_v = 1$), Equation (4.1) becomes

$$R_{TP} = 3R^2 - 2R^3 \qquad (4.2)$$

where
R_{TP} = TMR system with perfect voter reliability

Note that improvement in reliability of the TMR system over a single-unit/ module system is determined by the reliability of the single-unit/module system and the voter's reliability. When the voter is perfect, the reliability of the TMR system expressed by Equation (4.2) is only better than the single-unit/module system when the reliability of the single unit or module is higher than 0.5.

When the voter reliability is 0.9 (i.e., $R_v = 0.9$), the reliability of the TMR system is only marginally better than the reliability of a single-unit/ module system when the single-unit/module reliability is approximately between 0.833 and 0.667 [18]. Moreover, at $R_v = 0.8$, the reliability of the TMR system is always less than the reliability of a single-unit/module system.

4.5.1.1 TMR System Maximum Reliability with Perfect Voter

For the perfect voter, the reliability of the TMR system is expressed by Equation (4.2). Under this scenario, the ratio of R_{TP} to the reliability of a single unit, R, is expressed by [19]

$$\alpha = \frac{R_{TP}}{R}$$

$$= \frac{3R^2 - 2R^3}{R} \tag{4.3}$$

$$= 3R - 2R^2$$

By differentiating Equation (4.3) with respect to R and equating it to zero, we obtain

$$\frac{d\alpha}{dR} = 3R - 4R = 0 \tag{4.4}$$

Thus, from Equation (4.4), we get $R = 0.75$.

This simply means that the maximum values of the reliability improvement ratio, α, and the TMR system's reliability, R_{TP}, are respectively:

$$\alpha = 3(0.75) - 2(0.75)^2$$

$$= 1.125$$

and

$$R_{TP} = 3(0.75)^2 - 2(0.75)^3$$

$$= 0.8438$$

Example 4.1

Assume that the reliability of a TMR system with perfect voter is given by Equation (4.2). Determine the points where the single-unit and the TMR-system reliabilities are equal.

To determine the points, we equate the reliability, R, of a single unit with Equation (4.2) to get

$$R = R_{TP} = 3R^2 - 2R^3 \tag{4.5}$$

By rearranging Equation (4.5), we obtain

$$2R^2 - 3R + 1 = 0 \tag{4.6}$$

Equation (4.6) is a quadratic equation and its roots are

$$R = \frac{3 + \left[9 - 4(2)(1)\right]^{1/2}}{2(2)} = 1 \tag{4.7}$$

and

$$R = \frac{3 - \left[9 - 4(2)(1)\right]^{1/2}}{2(2)} = \frac{1}{2} \tag{4.8}$$

This means that the reliabilities of the single unit and the TMR system with perfect voter are equal at R = either 1 or ½. Furthermore, the reliability of the TMR system with perfect voter will only be higher than the single-unit reliability when the value of R is greater than 0.5.

4.5.1.2 TMR System Time-Dependent Reliability and Mean Time to Failure

With the aid of material presented in Chapter 3 and Equation (4.1), for constant failure rates of the TMR system units and the voter unit, the reliability of the TMR system with voter is given by [9, 20]

$$\begin{aligned} R_{TV}(t) &= \left[3e^{-2\lambda t} - 2e^{-3\lambda t}\right] e^{-\lambda_v t} \\ &= 3 - e^{-(2\lambda + \lambda_v)t} - 2e^{-(3\lambda + \lambda_v)t} \end{aligned} \tag{4.9}$$

where
$R_{TV}(t)$ = TMR system with voter reliability at time t
λ_v = voter unit constant failure rate
λ = module/unit constant failure rate

By integrating Equation (4.9) over the time interval from 0 to ∞, we get the following expression for the mean time to failure of the TMR system with voter [6, 9]:

$$\begin{aligned} MTTF_{TV} &= \int_0^\infty \left[3e^{-(2\lambda + \lambda_v)t} - 2e^{-(3\lambda + \lambda_v)t}\right] dt \\ &= \frac{3}{(2\lambda + \lambda_v)} - \frac{2}{(3\lambda + \lambda_v)} \end{aligned} \tag{4.10}$$

For the perfect voter (i.e., $\lambda_v = 0$), Equations (4.9) and (4.10) simplify, respectively, to

$$R_{TP}(t) = 3e^{-2\lambda t} - 2e^{-3\lambda t} \tag{4.11}$$

and

$$\text{MTTF}_{TP} = \frac{5}{6\lambda} \tag{4.12}$$

where
$R_{TP}(t)$ = reliability of TMR system with perfect voter at time t
MTTF_{TP} = mean time to failure for TMR system with perfect voter

Example 4.2

Assume that the constant failure rate of a module/unit belonging to a TMR system with voter is $\lambda = 0.0002$ failures per hour. Calculate the system reliability for an 800-hour mission if the voter constant failure rate is $\lambda_v = 0.0001$ failures per hour. In addition, calculate the system mean time to failure.

By substituting the given data values into Equation (4.9), we obtain

$$R_{TV}(800) = 3e^{-[2(0.0002)+0.0001](800)} - 2e^{-[3(0.0002)+0.0001](800)}$$

$$= 0.8685$$

Similarly, by substituting the given data values into Equation (4.10), we obtain

$$\text{MTTF}_{TV} = \frac{3}{[2(0.0002)+0.0001]} - \frac{2}{[3(0.0002)+0.0001]}$$

$$= 3{,}142.86 \text{ hours}$$

Thus, for the TMR system with voter, the system reliability (R_{TV}) and the mean time to failure (MTTF_{TV}) are 0.8685 and 3,142.86 hours, respectively.

Example 4.3

Repeat the Example 4.2 calculations for a TMR system having a perfect voter and comment on the end results. In this case, as the voter is perfect, its failure rate is equal to zero (i.e., $\lambda_v = 0$). Thus, by inserting the remaining data values into Equations (4.11) and (4.12), we get

$$R_{TP}(800) = 3e^{-2(0.0002)(800)} - 2e^{-3(0.0002)(800)}$$

$$= 0.9408$$

and

$$\text{MTTF}_{TP} = \frac{5}{6(0.0002)}$$

$$= 4{,}166.66 \text{ hours}$$

In this case, the TMR system with a perfect voter has a reliability (R_{TP}) of 0.9408 and a mean time to failure ($MTTF_{TP}$) of 4,166.66 hours. By comparing these results with those of Example 4.2, it can be observed that the perfect voter has improved the TMR system's reliability as well as its mean time to failure.

4.5.1.3 Reliability Analysis of TMR System with Perfect Voter and Repair

In the preceding discussion of a TMR system with and without perfect voter reliability analysis, all units/modules were considered nonrepairable. This is true in certain circumstances or applications, but in others it is possible to repair the failed modules/units. Thus, in this case, we consider that the units of the TMR system with perfect voter are repairable. More specifically, whenever any one of the TMR system units fails, it is repaired. However, when more than one unit fails, the TMR system is not repaired. The TMR system transition or state space diagram is shown in Figure 4.2

The Markov method presented in Chapter 3 is used to develop state probability equations for the state space diagram shown in Figure 4.2 [6, 9]. The following assumptions are associated with the TMR system model presented in Figure 4.2:

- All system units are active, and the system fails when two or more units fail.
- The system is composed of three identical and independent units with a perfect voter.
- Failure and repair rates of a unit are constant.
- A repaired unit is as good as new.
- The system is only repaired when one unit fails.
- The numerals in Figure 4.2 boxes denote specific system states.

The following symbols are associated with this model:

i = ith state of the TMR system shown in Figure 4.2 for $i = 0$ (all three units operating; TMR system up), $i = 1$ (one unit failed; TMR system up), $i = 2$ (two units failed; TMR system failed)

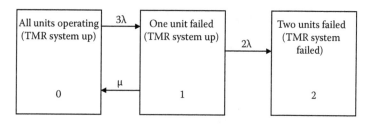

FIGURE 4.2
State space diagram for TMR system with perfect voter and repair.

$P_i(t)$ = probability that the TMR system is in state i at time t for i = 0,1,2
λ = module/unit constant failure rate
μ = module/unit constant repair rate

Using the Markov method described in Chapter 3 and in the literature [9, 21], we write down the following set of differential equations for Figure 4.2:

$$\frac{dP_0(t)}{dt} + 3\lambda P_0(t) = \mu P_1(t) \tag{4.13}$$

$$\frac{dP_1(t)}{dt} + (2\lambda + \mu)P_1(t) = 3\lambda P_0(t) \tag{4.14}$$

$$\frac{dP_2(t)}{dt} = 2\lambda P_1(t) \tag{4.15}$$

At time t = 0, $P_0(0)$ = 1 and $P_1(0)$ = $P_2(0)$ = 0

By solving Equations (4.13)–(4.15), we obtain the following expression for the reliability of a TMR system with perfect voter and repair [9]:

$$R_{TPR}(t) = P_0(t) + P_1(t)$$

$$= \frac{1}{y_1 - y_2}\left[(5\lambda + \mu)\left(e^{y_1 t} - e^{y_2 t}\right) + y_1 e^{y_1 t} - y_2 e^{y_2 t}\right] \tag{4.16}$$

where

$$y_1, y_2 = \left[-(5\lambda + \mu) \pm (\lambda^2 + \mu^2 + 10\lambda\mu)^{1/2}\right]/2 \tag{4.17}$$

By integrating Equation (4.16) over the time interval [0, ∞], we obtain

$$MTTF_{TPR} = \int_0^\infty R_{TPR}(t)\,dt$$

$$= \frac{5}{6\lambda} + \frac{\mu}{6\lambda^2} \tag{4.18}$$

where
$MTTF_{TPR}$ = mean time to failure for the TMR system with perfect voter and repair

Note that for μ = 0, Equation (4.18) becomes exactly the same as Equation (4.12).

Example 4.4

Assume that the constant failure rate of a module/unit of an independent TMR system, with perfect voter and repair, is 0.002 failures per hour. The module/unit repair rate is 0.4 repairs per hour. Calculate the mean time to failure for a TMR system with perfect voter and repair. Also, calculate the mean time to failure for a TMR system with perfect voter but no repair. Comment on the end results.

By substituting the given data values into Equation (4.18), we get

$$\text{MTTF}_{\text{TPR}} = \frac{5}{6(0.002)} + \frac{0.4}{6(0.002)^2}$$

$$= 17{,}083.3 \text{ hours}$$

Similarly, for no repair (i.e., $\mu = 0$), by substituting the given data value into Equation (4.12), we get

$$\text{MTTF}_{\text{TP}} = \frac{5}{6(0.002)}$$

$$= 416.6 \text{ hours}$$

This means that the introduction of repair has helped to increase the mean time to failure of the TMR system with perfect voter from 416.66 hours to 17,083.3 hours.

4.5.2 N-Modular Redundancy (NMR)

This is the general form of the TMR. More clearly, it contains N identical modules/units instead of only three units. The number N is any odd number and is expressed by $N = 2k + 1$. The NMR system will be operational or successful if at least $(k + 1)$ modules/units function normally. As the voter acts in series with the N-module system, the complete system fails whenever a voter failure occurs.

The reliability of the NMR system with independent modules/units is expressed by [9, 22]

$$R_{\text{NV}} = \left[\sum_{j=0}^{k} \binom{N}{j} R^{N-1}(1-R)^j \right] R_{\text{v}} \tag{4.19}$$

$$\binom{N}{j} = \frac{N!}{(N-j)!\,j!} \tag{4.20}$$

where
R_{NV} = NMR system with voter reliability
R = module/unit reliability
R_{v} = voter reliability

The time-dependent reliability analysis of an NMR system can be carried out in a manner similar to the TMR system reliability analysis. Information on additional redundancy schemes is available in the work of Nerber [16].

4.6 Reliability Analysis of Redundant Computer Systems with Common-Cause Failures

Over the years, the occurrence of common-cause failures has received increasing attention in the industrial sector because of the realization that the assumption of independent unit failures may be violated under real life conditions. A common-cause failure may simply be defined as any instance where multiple units fail due to a single cause [6, 9, 23].

Some of the main causes for the occurrence of common-cause failures in engineering systems are as follows [6, 9, 15]:

External catastrophe. This includes natural phenomena such as tornado, flood, earthquake, and fire. The occurrence of natural events such as these may lead to simultaneous failure of redundant units in a system.

Operations and maintenance errors. These errors may occur due to factors such as improper maintenance, carelessness, and incorrect calibration or adjustment by humans.

Common external power source. Redundant units directly or indirectly fed from the same power source may fail simultaneously due to the failure of the power source.

Common external environment. This includes items such as vibrations, temperature, dust, humidity, and moisture.

Common manufacturer. The redundant units procured from the same manufacturer may have the same design or fabrication errors. For example, fabrication errors may occur due to poor soldering, use of wrong material, wiring a circuit board backward, etc.

Equipment design deficiency. This is the result of oversights during the equipment/system design phase, e.g., failure to consider interdependence between mechanical and electrical parts or the subsystems of a redundant system.

Functional deficiency. This may occur due to inappropriate instrumentation or poorly designed protective action.

These causes clearly indicate that in order to predict the realistic reliability of redundant computer systems, one must take into consideration the occurrence of common-cause failures as well, since redundant computer systems can fail simultaneously or near simultaneously due to a common-cause failure.

The following subsections present two mathematical models representing redundant computer systems subject to common-cause failures. In both cases, the Markov method is used to develop equations.

4.6.1 Model I

This mathematical model represents two redundant computers with common-cause failures. At least one computer must operate normally for the system's successful operation. The system fails when both computers fail. Furthermore, both computers may fail due to a common-cause failure. The system state space diagram is shown in Figure 4.3 [6]. The numerals in the boxes and circles of this figure denote system states. The following assumptions are associated with this model:

- Both computers are active (at time $t = 0$) and identical.
- The failed computer is never repaired.
- Common-cause and other failures occur independently.
- A common-cause failure can only occur when both computers are operating normally.
- The computer failure rate is constant.
- The common-cause failure rate is constant.

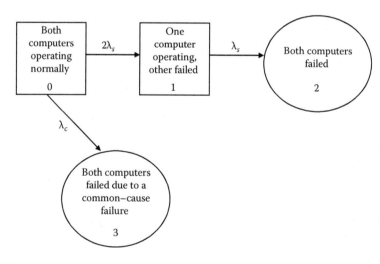

FIGURE 4.3

State space diagram for two redundant computers with common-cause failures.

The following symbols are associated with this model:

j = jth state of the redundant computer system: $j = 0$ (both computers operating normally), $j = 1$ (one computer operating, other failed), $j = 2$ (both computers failed), $j = 3$ (both computers failed due to a common-cause failure)

$P_j(t)$ = probability that the redundant computer system is in state j at time t for $j = 0,1,2,3$

λ_s = single computer constant failure rate

λ_c = redundant computer system constant common-cause failure rate

s = Laplace transform variable

By using the Markov method, we write down the following differential equations for Figure 4.3 [6, 9]:

$$\frac{dP_0(t)}{dt} + (2\lambda_s + \lambda_c)P_0(t) = 0 \tag{4.21}$$

$$\frac{dP_1(t)}{dt} + \lambda_s P_1(t) = 2\lambda_s P_0(t) \tag{4.22}$$

$$\frac{dP_2(t)}{dt} = \lambda_s P_1(t) \tag{4.23}$$

$$\frac{dP_3(t)}{dt} = \lambda_c P_0(t) \tag{4.24}$$

At time $t = 0$, $P_0(0) = 1$ and $P_1(0) = P_2(0) = P_3(0) = 0$

By solving Equations (4.21)–(4.24), we obtain the following equations:

$$P_0(s) = \frac{1}{s + 2\lambda_s + \lambda_c} \tag{4.25}$$

$$P_1(s) = \frac{2\lambda_s}{(s + 2\lambda_s + \lambda_c)(s + \lambda_s)} \tag{4.26}$$

$$P_2(s) = \frac{2\lambda_s^2}{s(s + 2\lambda_s + \lambda_c)(s + \lambda_s)} \tag{4.27}$$

$$P_3(s) = \frac{\lambda_c}{s(s + 2\lambda_s + \lambda_c)} \tag{4.28}$$

where

$P_j(s)$ = Laplace transform of the jth state probability for $j = 0,1,2,3$

By taking the inverse Laplace transforms of Equations (4.25)–(4.28), we get the following state probability equations:

$$P_0(t) = e^{-(2\lambda_s + \lambda_c)t} \tag{4.29}$$

$$P_1(t) = X\left[e^{-\lambda_s t} - e^{-(2\lambda_s + \lambda_c)t}\right] \tag{4.30}$$

where

$$X = \frac{2\lambda_s}{(2\lambda_s + \lambda_c - \lambda_s)} \tag{4.31}$$

$$P_2(t) = Y + X\left[\frac{\lambda_s}{2\lambda_s + \lambda_c}e^{-(2\lambda_s + \lambda_c)t} - e^{-\lambda_s t}\right] \tag{4.32}$$

where

$$Y = \frac{2\lambda_s}{(2\lambda_s + \lambda_c)} \tag{4.33}$$

$$P_3(t) = \frac{\lambda_c}{(2\lambda_s + \lambda_c)}\left[1 - e^{-(2\lambda_s + \lambda_c)t}\right] \tag{4.34}$$

The system reliability for two redundant computers with common-cause failures is given by

$$\begin{aligned}
R_{TC}(t) &= P_0(t) + P_1(t) \\
&= e^{-(2\lambda_s + \lambda_c)t} + X\left[e^{-\lambda_s t} - e^{-(2\lambda_s + \lambda_c)t}\right]
\end{aligned} \tag{4.35}$$

where
   $R_{TC}(t)$ = reliability of system comprising two redundant computers with common-cause failures at time t

The Laplace transform of system reliability for two redundant computers with common-cause failures is given by

$$\begin{aligned}
R_{TC}(s) &= P_0(s) + P_1(s) \\
&= \frac{1}{(s + 2\lambda_s + \lambda_c)} + \frac{2\lambda_s}{(s + 2\lambda_s + \lambda_c)(s + \lambda_s)}
\end{aligned} \tag{4.36}$$

Using Dhillon's work [6, 9], we express the system mean time to failure for two redundant computers with common-cause failures as follows:

$$\text{MTTF}_{TC} = \lim_{s \to 0} R_{TC}(s) \tag{4.37}$$

where
MTTF_{TC} = system mean time to failure for two redundant computers with common-cause failures

Using Equation (4.36) in Equation (4.37) yields

$$
\begin{aligned}
\text{MTTF}_{TC} &= \lim_{s \to 0} \left[\frac{1}{(s + 2\lambda_s + \lambda_c)} + \frac{2\lambda_s}{(s + 2\lambda_s + \lambda_c)(s + \lambda_s)} \right] \\
&= \frac{3}{(2\lambda_s + \lambda_c)}
\end{aligned}
\tag{4.38}
$$

Example 4.5

Assume that a system has two independent and identical computers with common-cause failures. At least one of the computers must operate normally for the system to operate successfully. The constant failure rate of a computer is 0.0005 failures per hour, and the constant common-cause failure rate of the redundant system is 0.00002 failures per hour. Calculate the system mean time to failure with and without common-cause failures. Comment on the end results.

By substituting the specified data values into Equation (4.38), we obtain the following value for the system mean time to failure with common-cause failures:

$$\text{MTTF}_{TC} = \frac{3}{2(0.0005) + 0.00002}$$

$$= 2{,}941.17 \text{ hours}$$

By setting $\lambda_c = 0$ (i.e., the common-cause failure rate is zero) and then substituting the remaining given data value into Equation (4.38), we obtain the following value for the system mean time to failure without common-cause failures:

$$\text{MTTF}_T = \frac{3}{2(0.0005)}$$

$$= 3{,}000 \text{ hours}$$

This means that the occurrence of common-cause failures has decreased the system mean time to failure from 3,000 hours to 2,941.17 hours.

4.6.2 Model II

This mathematical model is the same as Model I, but with one exception, i.e., whenever a computer fails, it is repaired back to its operational state. In order to handle this situation, the diagram in Figure 4.3 is modified as shown in Figure 4.4. This figure shows a new transition from state 1 to state 0. The symbol μ in Figure 4.4 denotes the constant repair rate of a computer (a repaired computer is considered as good as new). The symbols and assumptions used to develop equations for Model I are equally applicable to Model II.

By using the Markov method, we write down the following set of differential equations for Figure 4.4 [6, 9]:

$$\frac{dP_0(t)}{dt} + (2\lambda_s + \lambda_c)P_0(t) = \mu P_1(t) \tag{4.39}$$

$$\frac{dP_1(t)}{dt} + (\lambda_s + \mu)P_1(t) = 2\lambda_s P_0(t) \tag{4.40}$$

$$\frac{dP_2(t)}{dt} = \lambda_s P_1(t) \tag{4.41}$$

$$\frac{dP_3(t)}{dt} = \lambda_c P_0(t) \tag{4.42}$$

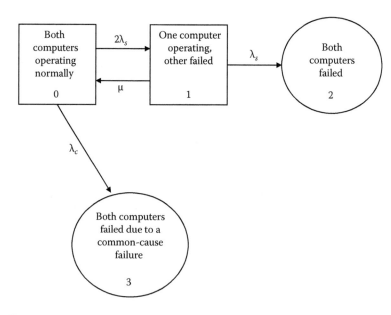

FIGURE 4.4
State space diagram for two redundant computers with common-cause failures and repair.

At time $t = 0$, $P_0(0) = 1$ and $P_1(0) = P_2(0) = P_3(0) = 0$

By solving Equations (4.39)–(4.42), we get the following equations:

$$P_0(s) = \frac{s + \lambda_s + \mu}{A} \qquad (4.43)$$

where

$$A = (s + 2\lambda_s + \lambda_c)(s + \lambda_s + \mu) - 2\lambda_s\mu \qquad (4.44)$$

$$P_1(s) = 2\lambda_s/A \qquad (4.45)$$

$$P_2(s) = 2\lambda_s^2/sA \qquad (4.46)$$

$$P_3(s) = \lambda_c (s + \lambda_s + \mu)/sA \qquad (4.47)$$

where
$P_j(s)$ = Laplace transform of the Figure 4.4 jth state probability for $j = 0,1,2,3$

By adding Equations (4.43) and (4.45), we obtain the Laplace transform of system reliability for two redundant computers with common-cause failures and repair as follows:

$$R_{TCR}(s) = P_0(s) + P_1(s)$$
$$= \frac{(s + \lambda_s + \mu)}{A} + \frac{2\lambda_s}{A} \qquad (4.48)$$

where
$R_{TCR}(s)$ = Laplace transform of system reliability for two redundant computers with common-cause failures and repair

Using Equation (4.48) in Equation (4.37) yields the mean time to failure for two redundant computers with common-cause failures and repair

$$MTTF_{TCR} = \lim_{s \to 0} \left[\frac{s + \lambda_s + 2\lambda_s + \mu}{A} \right]$$
$$= \frac{3\lambda_s + \mu}{(2\lambda_s^2 + \lambda_c\lambda_s + \lambda_c\mu)} \qquad (4.49)$$

where
$MTTF_{TCR}$ = mean time to failure for two redundant computers with common-cause failures and repair

Example 4.6

Assume that the computers in Example 4.5 are repaired and the constant repair rate of a failed computer is 0.04 repairs per hour. Calculate the repairable system's mean time to failure with and without common-cause failures. Comment on the end results.

By substituting the given data values into Equation (4.49), we obtain the following value for the repairable system's mean time to failure with common-cause failures:

$$\text{MTTF}_{TCR} = \frac{3(0.0005) + 0.04}{2(0.0005)^2 + (0.00002)(0.0005) + (0.00002)(0.04)}$$

$$= 31{,}679.38 \text{ hours}$$

By setting $\lambda_c = 0$ (i.e., the common-cause failure rate is zero) and then substituting the remaining given data values into Equation (4.49), we get the following value for the repairable system's mean time to failure without common-cause failures:

$$\text{MTTF}_{TR} = \frac{3(0.0005) + 0.04}{2(0.0005)^2}$$

$$= 83{,}000 \text{ hours}$$

This means that the occurrence of common-cause failures has decreased the repairable system's mean time to failure from 83,000 hours to 31,679.38 hours. Furthermore, in comparison to the end results of Example 4.5, the repair has helped to increase the system mean time to failure with common-cause failures from 2,941.17 hours to 31,679.38 hours and the system mean time to failure without common-cause failures from 3,000 hours to 83,000 hours.

Problems

1. Compare hardware reliability with software reliability.
2. Discuss the major sources of computer failures.
3. Discuss the main issues in computer system reliability.
4. Describe the following two classifications of computer faults:
   a. Transient faults
   b. Permanent faults
5. Define the following terms:
   a. Computation reliability
   b. Mean computation before failure
   c. Computation availability
   d. Computation threshold

6. What is triple modular redundancy (TMR)?

7. Assume that the constant failure rate of a unit belonging to a TMR system with voter is $\lambda = 0.0007$ failures per hour. Calculate the system reliability for a 200-hour mission if the voter constant failure rate is $\lambda_v = 0.0003$ failures per hour. In addition, calculate the system mean time to failure.

8. Prove Equation (4.18) by using Equations (4.16) and (4.17).

9. Compare N-Modular Redundancy (NMR) with Triple Modular Redundancy (TMR).

10. Define a common-cause failure. What are the main causes for the occurrence of common-cause failures in engineering systems?

References

1. Shannon, C. E. (1948). A mathematical theory of communications. *Bell System Tech. J.*, *27*, 379–423, 623–656.
2. Hamming, W. R. (1950). Error detecting and error correcting codes. *Bell System Tech. J.*, *29*, 147–160.
3. Moore, E. F., & Shannon, C. E. (1956). Reliable circuits using less reliable relays. *J. Franklin Inst.*, *262*, 191–208.
4. Von Neumann, J. (1956). Probabilistic logics and the synthesis of reliable organisms from reliable components. In C. E. Shannon & J. McCarthy (Eds.), *Automata Studies* (pp. 43–98). Princeton, NJ: Princeton University Press.
5. Dhillon, B. S., & Ugwu, K. I. (1986). Bibliography of literature on computer hardware reliability. *Microelectronics and Reliability*, *26*, 99–122.
6. Dhillon, B. S. (1987). *Reliability in computer system design*. Norwood, NJ: Ablex Publishing.
7. Kline, M. B. (1980). Software and hardware reliability and maintainability: What are the differences? *Proceedings of the Annual Reliability and Maintainability Symposium*, 179–185.
8. Ireson, W. G., Coombs, C. F., Jr., & Moss, R. Y. (1996). *Handbook of reliability engineering and management*. New York, NY: McGraw Hill.
9. Dhillon, B. S. (1999). *Design reliability: Fundamentals and applications*. Boca Raton, FL: CRC Press.
10. Dhillon, B. S. (1983). *Reliability engineering in systems design and operation*. New York, NY: Van Nostrand Reinhold.
11. Yourdon, E. (1972). The causes of system failures: Part II. *Modern Data*, *5*, 36–40.
12. Yourdon, E. (1972). The causes of system failures: Part III. *Modern Data*, *5*, 50–56.
13. Goldberg, J. (1975). A survey of the design and analysis of fault-tolerant computers. In R. E. Barlow, J. B. Fussell, and N. D. Singpurwalla (Eds.), *Reliability and fault tree analysis* (pp. 667–685). Philadelphia, PA: Society for Industrial and Applied Mathematics.

14. Avizienis, A. (1977). Fault-tolerant computing: Progress, problems, and prospectus. *Proceedings of the International Federation for Information Processing (IFIP) Congress*, 405–420.
15. Dhillon, B. S., & Singh, C. (1981). *Engineering reliability: New techniques and applications*. New York, NY: John Wiley and Sons.
16. Nerber, P. O. (1965, March). Power-off time impact on reliability estimates. *IEEE Int. Conv. Record*, Part 10, 1–8.
17. Mathur, F. P., & Avizienis, A. (1970). Reliability analysis and architecture of a hybrid redundant digital system: Generalized triple modular redundancy with self-repair. *Proceedings of the AFIPS Conference*, 375–387.
18. Pecht, M. (Ed.). (1995). *Product reliability, maintainability, and supportability handbook*. Boca Raton, FL: CRC Press.
19. Shooman, M. L. (1994). Fault tolerant computing. *Annual Reliability and Maintainability Symposium Tutorial Notes*, 1–25.
20. Dunn, R., & Ullman, R. (1982). *Quality assurance for computer software*. New York, NY: McGraw Hill.
21. Shooman, M. L. (1968). *Probabilistic reliability: An engineering approach*. New York, NY: McGraw Hill.
22. Shooman, M. L. (2002). *Reliability of computer systems and networks: Fault tolerance, analysis, and design*. New York, NY: John Wiley and Sons.
23. Dhillon, B. S., & Anude, D. C. (1994). Common-cause failures in engineering systems: A review. *International Journal of Reliability, Quality, and Safety Engineering, 1*, 103–129.

5

Software Reliability Assessment and Improvement Methods

5.1 Introduction

Over the years, the percentage of the total computer cost spent on the software component has changed dramatically from the days of first-generation computers. For example, in 1955 the software portion (i.e., including software maintenance) accounted for about 20% of the total computer cost, and in 1985, the cost of the software portion had increased to around 90% [1, 2]. As the result of this remarkable change, many software reliability assessment and improvement methods/models have been developed over the years.

It appears that the serious effort on software reliability started at Bell Laboratories in the early 1960s [3]. A clear evidence of such effort is a histogram of monthly problems in regard to switching system software. In the latter part of the 1960s, approaches for formal validation of software programs were considered [4], and Markov birth-death models directly or indirectly concerned with software reliability were proposed [5].

Nowadays, there are many methods concerning software reliability assessment and improvement. This chapter presents a number of such methods extracted from the published literature that are considered to be the most useful.

5.2 Software Reliability Assessment Methods

There are many qualitative and quantitative methods that can be used to assess software reliability. These methods may be grouped under three classifications as shown in Figure 5.1 [6]. The classifications shown in Figure 5.1 are software metrics, analytical methods, and software reliability models. Each of these classifications is described separately in the following subsections.

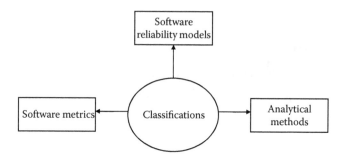

FIGURE 5.1
Classifications of software reliability assessment methods.

5.2.1 Software Metrics

These metrics may simply be described as quantitative indicators of the degree to which a software process/item possesses a stated attribute. Frequently, software metrics are used for determining the status of a trend in a software development process and for determining the risk of going from one phase to another. Two software metrics considered useful to assess, directly or indirectly, software reliability are presented in the following subsections.

5.2.1.1 Metric I: Code and Unit Test Phase Measure

This metric is concerned with assessing software reliability during the code and unit test phase. The metric is defined by [2, 6, 7]

$$CDRC = \sum_{j=1}^{n} CDF_j / SLC \qquad (5.1)$$

where
    CDRC = cumulative defect ratio for code
    n = number of reviews
    SLC = total number of source lines of code reviewed, expressed in thousands
    CDF_j = number of unique defects, at or above a stated severity level, found in the jth code review

5.2.1.2 Metric II: Design Phase Measure

This metric is concerned with determining the reliability growth during the design phase. The metric/measure required establishing appropriate defect severity classifications and possesses some ambiguity, since its low value may mean either a good product or a poor review process. The metric is defined by [2, 6, 7]

$$CDRD = \sum_{j=1}^{n} DDF_j / SLD \qquad (5.2)$$

where
   CDRD = cumulative defect ratio for design
   n = number of reviews
   SLD = total number of source lines of design statement in the design phase, expressed in thousands
   DDF_j = number of unique defects, at or above a stated severity level, found in the jth design review

Additional information on this metric is available in the literature [2, 6, 7].

5.2.2 Analytical Methods

There are a number of analytical methods that can be used to assess software reliability. Two of these methods are fault tree analysis (FTA) and failure modes and effect analysis (FMEA). Both of these methods are commonly used to assess reliability of hardware, and they can equally be used to assess reliability of software as well. FTA and FMEA are described in detail in Chapter 3.

5.2.3 Software Reliability Models

There are many software reliability models [8–10]. All these models may be grouped under four classifications as follows [2, 6, 7]:

   Classification I: Fault seeding. This classification includes those software reliability models that determine the number of faults in the program at zero time via seeding of extraneous faults. Two main assumptions associated with the models belonging to Classification I are as follows:
   1. The seeded faults are distributed randomly in the software program under consideration.
   2. Indigenous and seeded faults have equal probability of detection.
      An example of the model belonging to this classification is the Mills model [11].
   Classification II: Times between failures. This classification includes those software reliability models that provide the time between

failure estimations. The following four main assumptions are associated with the models belonging to this classification:

1. Independent embedded faults
2. The correction process does not introduce faults
3. Independent times between failures
4. Equal probability of exposure of each fault

Two examples of the models belonging to Classification II are the Shick and Wolverton model [10] and the Jelinski and Moranda model [12].

Classification III: Input domain based. This classification includes those software reliability models that determine the software/program reliability under the condition that the test cases are sampled randomly from a given operational distribution of inputs to the software/program. Three main assumptions associated with models belonging to Classification III are as follows:

1. Given input profile distribution
2. Inputs chosen randomly
3. Input domain can be separated into equivalence groups

Two examples of the models belonging to this classification are the Nelson model [13] and the Ramamoorthy and Bastani model [14].

Classification IV: Failure count. This classification includes those software reliability models that count the number of faults/failures occurring in specified time intervals. The following three main assumptions are associated with the models belonging to this classification:

1. Independent test intervals
2. Independent faults found during non-overlapping time intervals
3. Homogeneously distributed testing during intervals

Two examples of the models belonging to Classification IV are the Shooman model [15] and the Musa model [16].

Some of the software reliability-related models are presented in the following subsections.

5.2.3.1 Air Force Model

This model was developed at the U.S. Air Force Rome Laboratory to predict software reliability during the initial phases of the software life cycle [17]. The model predicts fault density, a metric that can subsequently be transformed to other reliability measures, including failure rates [18]. The factors related to fault density at the initial phases are program size, development environment, language type, nature of application, anomaly management, traceability, standards review results, complexity, quality review results, modularity, and extent of reuse. Thus, the initial fault density is defined by [17, 18]

$$\text{IFD} = \prod_{i=1}^{11} f_i \qquad (5.3)$$

where
 IFD = initial fault density
 f_i = ith factor related to fault density at the initial phases: for $i = 1$ (program size), $i = 2$ (development environment), $i = 3$ (language type), $i = 4$ (nature of application), $i = 5$ (anomaly management), $i = 6$ (traceability), $i = 7$ (standards review results), $i = 8$ (complexity), $i = 9$ (quality review results), $i = 10$ (modularity), and $i = 11$ (extent of reuse)

The initial failure is defined by

$$\lambda_i = \alpha\beta\theta = \alpha\beta(\mu\text{IFD}) \qquad (5.4)$$

where
 λ_i = initial failure rate
 α = linear execution frequency of the program
 β = fault expose ratio ($1.4 \times 10^{-7} \geq \beta \leq 10.6 \times 10^{-7}$)
 μ = number of lines of source code

The linear execution frequency of the program, α, is defined by

$$\alpha = \frac{m}{n} \qquad (5.5)$$

where
 n = total number of object instructions in the program
 m = average instruction rate

The total number of object instructions in the program, n, is expressed by

$$n = (\text{NSI})(R_c) \qquad (5.6)$$

where
 NSI = total number of source instructions
 R_c = code expansion ratio (Note that this is the ratio of machine instructions to source instructions and, generally, its mean value is taken as 4.)

The number of inherent faults is defined by

$$F_{in} = (\text{IFD})\mu \qquad (5.7)$$

where
 F_{in} = number of inherent faults

By substituting Equations (5.5)–(5.7) into Equation (5.4), we get

$$\lambda_i = m\beta F_{in}/(NSI)(R_c) \tag{5.8}$$

5.2.3.2 Musa Model

The basis for this model is the premise that reliability assessments in the time domain can only be based upon actual execution time, as opposed to elapsed or calendar time. The reason for this is that only during the execution process does a software program becomes exposed to failure-provoking stress. Some of the main assumptions associated with this model are as follows [2, 16]:

- Failure intervals follow a Poisson distribution and are statistically independent.
- Execution time between failures is piecewise exponentially distributed, and failure rate is proportional to the remaining defects.

The net number of corrected faults is expressed by [2, 7, 16]

$$n = N\,[1 - \exp(-ct/NMT_s)] \tag{5.9}$$

where
n = net number of corrected faults
t = time
N = initial number of faults
C = testing compression factor expressed as the average ratio of detection rate of failures during test to the rate during normal use of the software program under consideration
MT_s = mean time to failure at the start of the test

Mean time to failure increases exponentially with execution time and is defined by

$$MTTF = MT_s\,\exp(ct/NMT_s) \tag{5.10}$$

Thus, the reliability at operational time t is

$$R(t) = \exp(-t/MTTF) \tag{5.11}$$

From the above relationships, we get the number of failures that must occur for improving mean time to failure from, say, $MTTF_1$ to $MTTF_2$ [19]:

$$\Delta n = NMT_s\left[\frac{1}{MTTF_1} - \frac{1}{MTTF_2}\right] \tag{5.12}$$

The additional execution time needed to experience Δn is given by

$$\Delta t = \left[\frac{NMT_s}{c}\right] \ln\left(\frac{MTTF_2}{MTTF_1}\right) \tag{5.13}$$

Example 5.1

Assume that a newly developed software program is estimated to have around 200 errors. Also, at the start of the testing process, the recorded mean time to failure is 3 hours.

Determine the amount of test time needed to reduce the remaining errors to 10 if the value of the testing compression factor is 4. Also, calculate reliability over a 200-hour operational period.

By inserting the specified data values into Equation (5.12), we obtain

$$(200-10) = (200)(3)\left[\frac{1}{3} - \frac{1}{MTTF_2}\right] \tag{5.14}$$

Rearranging Equation (5.14) yields

$$MTTF_2 = 60 \text{ h}$$

By substituting the above result and the other given data values into Equation (5.13), we obtain

$$\Delta t = \left[\frac{(200)(3)}{4}\right] \ln\left(\frac{60}{3}\right)$$

$$= 449.36 \text{ h}$$

Thus, for the specified and calculated values from Equation (5.11), we obtain

$$R(200) = \exp\left(-\frac{200}{60}\right)$$

$$= 0.0357$$

Thus, the needed testing time is 449.36 hours, and the reliability of the software program for the given operational time period is 0.0357.

5.2.3.3 *Mills Model*

The basis for this model is that an assessment of the faults remaining in a software program can be made through a seeding process that assumes a homogeneous distribution of a representative category of faults. Prior to the initiation of the seeding process, a fault analysis is needed to determine the expected types of faults in the code and their relative frequency of occurrence.

An identification of unseeded and seeded faults is made during the review or testing process, and the discovery of indigenous and seeded faults permits an assessment of remaining faults for the type of fault under consideration. However, it is to be noted with care that the value of this measure can only be calculated if the seeded faults are found.

The maximum likelihood of the unseeded faults is defined by [6, 11]

$$M_{uf} = [M_{sf}n_{uf}]/n_{sf} \tag{5.15}$$

where
M_{uf} = maximum likelihood of the unseeded faults
n_{sf} = number of seeded faults found
n_{uf} = number of unseeded faults uncovered
M_{sf} = number of seeded faults

Thus, the number of unseeded faults still remaining in a software program under consideration is given by

$$M = M_{uf} - n_{uf} \tag{5.16}$$

where
M = number of unseeded faults still remaining in a software program under consideration

Example 5.2

Assume that a software program was seeded with 40 faults and that, during the testing process, 80 faults of the same type were discovered. The breakdowns of the faults uncovered were 60 unseeded faults and 20 seeded faults. Calculate the number of unseeded faults still remaining in the software program.

By inserting the specified data values into Equation (5.15), we obtain

$$M_{uf} = (40)(60)/20$$

$$= 120 \text{ faults}$$

By substituting the above resulting value and the other given data value into Equation (5.16), we obtain

$$M = 120 - 60 = 60 \text{ faults}$$

This means that 60 unseeded faults still remain in the software program.

5.2.3.4 Power Model

This model was proposed by Duane [20] for hardware reliability, but similar behavior has been observed for software products. Thus, from time to time, the model is referred to as the Duane model. In this model, the mean value function, $m(t)$, for the cumulative number of failures by time t is taken as a power of t, i.e.,

$$m(t) = \beta t^n, \quad \text{for} \quad n > 0, \beta > 0 \tag{5.17}$$

For $n = 1$, we get the homogeneous Poisson process model. Thus, the main assumption associated with the Power model is that the cumulative number of failures by time t, $m(t)$, follows a Poisson process with a value function represented by Equation (5.17). For the implementation of the model, the data requirement could be either of the following [18]:

- Actual times of software program failed (i.e., $t_1, t_2, t_3, \ldots, t_k$)
- Elapsed times between failures (i.e., $x_1, x_2, x_3, \ldots, x_k$ where $x_j = t_j - t_{j-1}$ and $t_0 = 0$)

If the time, for which the software program was under observation is T, then we can simply write

$$\frac{m(T)}{T} = \frac{\beta T^n}{T} = \frac{\text{Expected number of faults by } T}{\text{Total testing time, } T} \tag{5.18}$$

By taking the natural logarithms of Equation (5.18), we obtain

$$Y = \ln \beta + (n - 1)\ln T \tag{5.19}$$

It is to be noted that Equation (5.19) plots as a straight line, and it is the form fitted to a given data set.

By differentiating Equation (5.17) with respect to time t, we obtain the following equation for the failure intensity function:

$$\frac{dm(t)}{dt} = \beta n t^{n-1} = \lambda(t) \tag{5.20}$$

For $n > 1$, Equation (5.20) is strictly increasing; thus there can be no growth in reliability [18].

With the aid of the maximum likelihood estimation method given by Dhillon [2], we obtain [21]

$$\hat{\beta} = k / t_k^{\hat{n}} \tag{5.21}$$

and

$$\hat{n} = k \Bigg/ \sum_{j=1}^{k-1} \ln(t_k/t_j) \qquad (5.22)$$

Additional information on this model is available in the literature [18, 20, 21].

5.2.3.5 Shooman Model

This is a quite useful model; it does not require fault collection during the debugging process continuously; and it can be used for software of all sizes. The model is subjected to the following main assumptions [2, 7]:

- The number of errors in the software program remains constant at the start of the integration process, and they decrease directly as errors are eradicated.
- The debugging process does not introduce new errors.
- Total machine instructions remain constant.
- The residual errors are obtained by subtracting the total number of cumulative corrected errors from the total number of errors initially present.
- The hazard function is proportional to the remaining or residual software errors.

The hazard rate of the model is defined by

$$\lambda(t) = \theta F_m(y) \qquad (5.23)$$

where
t = system operating time
$\lambda(t)$ = hazard rate
y = debugging or elapsed time since the start of system integration
$F_m(y)$ = number of faults still remaining in the software program at time y

In turn, $F_m(y)$ is expressed by

$$F_m(y) = \frac{F_0}{\gamma} - F_u(y) \qquad (5.24)$$

where
γ = number of machine language instructions
$F_u(y)$ = cumulative number of faults corrected in interval y
F_0 = total number of initial faults at time $y = 0$

By substituting Equation (5.24) into Equation (5.23), we get

$$\lambda(t) = \theta \left[\frac{F_0}{\gamma} - F_u(y) \right] \tag{5.25}$$

As per Dhillon [2, 7], in reliability theory, the reliability of an item at time t is defined by

$$R(t) = \exp \left[-\int_0^t \lambda(t)\,dt \right] \tag{5.26}$$

Thus, by substituting Equation (5.25) into Equation (5.26), we obtain

$$R(t) = \exp \left[-\theta \left\{ \frac{F_0}{\gamma} - F_u(y) \right\} t \right] \tag{5.27}$$

By integrating Equation (5.27) over the time interval [0, ∞], we obtain the following expression for mean time to failure:

$$\text{MTTF} = \int_0^\infty \exp \left[-\theta \left\{ \frac{F_0}{\gamma} - F_u(y) \right\} t \right] dt \tag{5.28}$$

$$= 1 \Big/ \theta \left\{ \frac{F_0}{\gamma} - F_u(y) \right\} \tag{5.29}$$

The constants, θ and F_0, can be estimated by using the maximum-likelihood estimation method given by Dhillon [2].

5.3 Software Reliability Improvement Methods

Over the years, many methods have been developed to improve the reliability of software products. The categories of these methods include the following classifications [2, 7]:

- Fault-tolerant software design methods
- Reliable software design methods
- Testing
- Formal methods

Each of these four classifications is described separately in the following subsections.

5.3.1 Fault-Tolerant Software Design Methods

Three methods belonging to this classification are N-version programming, recovery-block design, and consensus recovery block.

N-version programming. In this case, many versions of the software are developed independently and N programs are executed simultaneously or in parallel. At the end, the comparisons are made of the resulting outputs, and when at least programs have the same output, it is considered as the correct result and the process continues. Finally, it is to be noted that this method is not suitable in situations where multiple correct outputs can be generated; also, the method discriminates correct solutions or results affected by rounding errors.

Recovery-block design. This method is concerned with developing versions of software under the same set of specifications/requirements by assuming that the pieces of software being developed are quite independent, and thus the likelihood of their simultaneous failure is negligible. The software versions are ordered from the least reliable to the most reliable. The software considered most reliable is executed first, and after the acceptance test run, if its resulting output is rejected, only then is a recovery block entered. The recovery is performed in the following three steps:

Step 1: Recover output.

Step 2: Execute software considered most reliable and restore input.

Step 3: Submit the output to the same acceptance test as before; in the event of its rejection, the next inline recovery block is entered and the process continues.

Consensus recovery block. This method combines attributes of the preceding two methods; thus it attempts to discard their weaknesses. Nonetheless, this method calls for the development of N-versions of a software as well as an acceptance-test voting procedure. The reliability factor is used to rank the different versions of the software after execution of all versions of the software in question. The resulting outputs are submitted for voting. In the event of no agreement, the order of reliability is employed by successively submitting each output to the acceptance test. As soon as one of the resulting outputs passes the test, the process terminates and the software in question continues with its operation.

All in all, it is to be noted that this method is generally considered more reliable than the preceding two methods (i.e., N-version programming and recovery-block design).

5.3.2 Reliable Software Design Methods

There are a number of methods that can be quite useful to programmers seeking to systematically derive the software design from its detailed specification [2, 22]. Three of these methods are as follows [2, 22]:

Top-down programming or design. This is a decomposition process concerned with (a) directing attention to the program control structure or to the program flow of control and (b) the production of code at a later stage. The top-down programming starts with a module representing the entire program, and it breaks down the module into a number of subroutines. In turn, all subroutines are broken down further, and this process continues until all the broken-down components or elements are easy and straightforward to comprehend and work with.

There are many advantages of top-down programming, including software in a more readable form, increased confidence in the software, reduction in the cost of testing, lower cost of software maintenance, and better quality software.

Structured programming. The direct aim of this method is the clarity of the design. More specifically, the method calls for easy-to-comprehend design and a need for developing a program structure in which the key software elements and their interrelationships can easily be identified. Some of the easy-to-use rules associated with structured programming include: the restricted use of GO TO statements in modules; using a language and compiler with structured programming capabilities for the implementation purpose; ensuring that all statements, including subroutine calls, are commented; and ensuring that each module has only one entry and one exit [23].

Finally, the structured-programming method has several advantages. It is a useful method in localizing an error; it is an effective tool in understanding the program design by the designers and others; it increases the productivity of programmers in regard to instructions coded per worker-hour; and it maximizes the amount of code that can be reused in redesign work.

Data structure design. This method is also known as Jackson Structured Programming and is inclined toward the development of a detailed program design. The input files' structure drives the methodology and the resulting program design, and from these very files software is read and, by the output files' structure, it is generated. Using this method, an effective software design reproduces the data structure in the output/input files.

It may be added that currently this is probably the most systematically available design technique. Furthermore, it is simple, consistent (e.g., two different programmers will end up writing the same program), anti-inspirational, and teachable. In contrast, the main drawback of the method is that it is not applicable to scientifically based programs because the contents of these programs are essentially algorithmic.

5.3.3 Testing

There are many methods that belong to this classification, and testing itself may simply be described as the process of executing a software program to find errors. Some of the testing methods are described here [24–26]:

Top-down testing. This testing method is concerned with integrating and testing the software program from the very top end to the very bottom end, and in the program structure, the very top module is the only module that is unit tested in isolation. At the final conclusion of the top module testing, this module calls all the remaining modules one by one by merging with it. In turn, each and every combination is tested, and the process continues until the completion of combining and testing all of the involved modules [26].

The benefits of the top-down testing method include easy representation of test cases after adding the input/output functions, early demonstrations because of the early skeletal program, and efficiency in locating problem areas at the higher program levels.

In contrast, the drawbacks of the top-down testing method include the following: Stub modules may be complex; stub modules must be written; the representations of test cases in stubs can be quite difficult prior to adding the input/output functions; and test conditions could be difficult or impossible to create [6].

Module testing. This testing method is concerned with testing each and every module, generally in isolation from the rest of the system, but completely subject to the type of environments to be experienced in the system. Normally, a general module test program is developed rather than writing a completely new program for each and every module to be tested. All in all, it is advisable to perform module testing thoroughly. Additional information on this method is available in the literature [2, 27].

Bottom-up testing. This testing method is concerned with integrating and testing the complete software program from the very bottom end to the very top end, and the terminal modules are completely tested in isolation. These modules do not call any other modules, and at the end of the terminal module testing, the modules that directly call the tested modules are the very ones in line for testing. Note that the modules are not tested in isolation; in fact, the testing is performed together with the previously tested lower level modules. This process continues until reaching the very top of the program.

Some of the main advantages of the bottom-up testing approach are the ease in observing test results, ease in increasing test conditions, and efficiency in finding problem areas at the lower levels of the program. In contrast, the main drawbacks of the bottom-up testing approach are that the program does not exist as an entity until the inclusion of the final module and the driver modules must be produced.

5.3.4 Formal Methods

The formal methods include formal specification methods as well as formal verification [6, 28]. A detailed specification of the software-related requirements using a formal language is contained in the formal specification methods. More clearly, the traditional programming's natural language is not utilized; instead, a "formal" substitute having exactly defined syntax and semantics is used. Mathematics is a typical example of such a language. The formal specification methods are quite helpful in enforcing clarity in requirement definitions, and they leave very little room for misinterpretation and ambiguity, which is generally a concern when making use of a natural language.

The formal verification uses inference on the formal specification to demonstrate that the software satisfies its set objectives [29]. Finally, note that the formal methods are quite complex and thus are not properly understood by all of the involved professionals because of the need for considerable training.

Problems

1. Write an essay on software reliability assessment and improvement methods.
2. What are the classifications of software reliability assessment methods?
3. Discuss the following two classifications of software reliability models:
   a. Fault seeding
   b. Failure count
4. Compare software reliability models' classification "times between failures" with the software reliability models' classification "input domain based."
5. Discuss the following two software reliability models:
   a. Air Force model
   b. Mills model

6. Assume that a newly developed software program is estimated to have about 150 errors. Also, at the start of the testing process, the recorded mean time to failure is 4 hours. Calculate the amount of test time required to reduce the remaining errors to 12, if the value of the testing compression factor is 3. Also, calculate reliability over a 100-hour operational period.

7. Assume that a software program was seeded with 50 faults and, during testing, 100 faults of the same type were found. The breakdowns of the faults found were 55 unseeded faults and 45 seeded faults. Calculate the number of unseeded faults still remaining in the software program.

8. Discuss the following two fault-tolerant software design methods:
   a. N-version programming
   b. Recovery-block design

9. What are the advantages of structured programming?

10. Describe the following two testing methods:
   a. Top-down testing
   b. Bottom-up testing

References

1. Keene, S. J. (1992). Software reliability concepts. *Annual Reliability and Maintainability Symposium Tutorial Notes*, 1–21.
2. Dhillon, B. S. (1999). *Design reliability: Fundamentals and applications*. Boca Raton, FL: CRC Press.
3. Haugk, G., Tsiang, S. H., & Zimmerman, L. (1964). System testing of the no. 1 electronic switching system. *Bell System Tech. Journal, 43*, 2575–2592.
4. Floyd, R. W. (1967). Assigning meanings to program. *Math. Aspects Comp. Sci., 19*, 19–32.
5. Hudson, G. R. (1967). *Programming errors as a birth-and-death process* (Report No. SP-3011). Santa Monica, CA: System Development Corp.
6. Pecht, M. (Ed.). (1995). *Product reliability, maintainability, and supportability handbook*. Boca Raton, FL: CRC Press.
7. Dhillon, B. S. (1987). *Reliability in computer system design*. Norwood, NJ: Ablex Publishing.
8. Sukert, A. N. (1977). An investigation of software reliability models. *Proceedings of the Annual Reliability and Maintainability Symposium*, 478–484.
9. Musa, J. D., Iannino, A., & Okumoto, K. (1987). *Software reliability*. New York, NY: McGraw Hill.
10. Schick, G. J., & Wolverton, R. W. (1978). An analysis of competing software reliability models. *IEEE Transactions on Software Engineering, 4*, 104–120.

11. Mills, H. D. (1972). *On the statistical validation of computer programs* (Report No. 72-6015). Gaithersburg, MD: IBM Federal Systems Division.
12. Jelinski, Z., & Moranda, P. B. (1972). Software reliability research. In *Proceedings of the Statistical Methods for the Evaluation of Computer System Performance* (pp. 465–484). New York, NY: Academic Press.
13. Nelson, E. (1978). Estimating software reliability from test data. *Microelectronics and Reliability, 17*, 67–75.
14. Ramamoorthy, C. V., & Bastani, F. B. (1982). Software reliability: Status and perspectives. *IEEE Transactions on Software Engineering, 8*, 354–371.
15. Shooman, M. L. (1975). Software reliability measurement and models. *Proceedings of the Annual Reliability and Maintainability Symposium, 1975*, 312–327.
16. Musa, J. D. (1975). A theory of software reliability and its applications. *IEEE Transactions on Software Engineering, 1*, 312–327.
17. Rome Air Development Center. (1992). *Methodology for software reliability prediction and assessment* (Report No. RL-TR-92-52) (Vols. 1 and 2). Rome, NY: Rome Air Development Center, Griffiss Air Force Base.
18. Lyu, M. R. (Ed.). (1996). *Handbook of software reliability engineering*. New York, NY: McGraw Hill.
19. Dunn, R., & Ullman, R. (1982). *Quality assurance for computer software*. New York, NY: McGraw Hill.
20. Duane, J. T. (1964). Learning curve approach to reliability monitoring. *IEEE Transactions on Aerospace, 2*, 563–566.
21. Crow, L. H. (1974). Reliability analysis for complex repairable systems. In F. Proschan & R. J. Serfling (Eds.), *Reliability and biometry* (pp. 379–410). Philadelphia, PA: Society for Industrial and Applied Mathematics (SIAM).
22. Bell, D., Morrey, L., & Pugh, J. (1992). *Software engineering: A programming approach*. London: Prentice-Hall.
23. Wang, R. S. (1980). Program with measurable structure. *Proceedings of the American Society for Quality Control Conference*, 389–396.
24. Beizer, B. (1984). *Software system testing and quality assurance*. New York, NY: Van Nostrand Reinhold.
25. Myers, G. J. (1979). *The art of software testing*. New York, NY: John Wiley and Sons.
26. Myers, G. J. (1976). *Software reliability: Principles and practices*. New York, NY: John Wiley and Sons.
27. Kopetz, H. (1979). *Software reliability*. London: Macmillan.
28. Neuhold, E. J., & Paul, M. (1991). Formal description of programming concepts. *Proc. of the Int. Fed. Info. Process. (IFIP) Conf.*, 310–315.
29. Galton, A. (1992). Logic as a formal method. *Comp. J., 35*(5), 213–218.

6

Software Quality

6.1 Introduction

Each year billions of dollars are spent to develop various types of computer software, and the quality of computer software has become a critical factor in the effective functioning of a computer. Nowadays, computers are widely used in various types of products or systems. Furthermore, it can be added that no product or system is of greater quality than the quality of its parts or elements, and if any of the parts or elements is a computer, then the quality of the software controlling that computer will definitely affect the overall product quality.

Software quality may simply be defined as the fitness for use of the software item/product [1, 2]. Thus, the main objective of a quality assurance program is to ensure that the final software items/products are of good quality through appropriately planned and systematic actions for determining, achieving, and maintaining that quality [2, 3]. This chapter presents various important aspects of software quality.

6.2 Software Quality Factors and Their Classifications

A large number of issues concerning various attributes of computer software and its maintenance and use, as outlined in software requirement documents, may be classified under content groups called quality factors. In turn, all software quality factors can be grouped under the following three classifications [4]:

- Product revision factors
- Product operation factors
- Product transition factors

These classifications are described in the following subsections.

6.2.1 Product Revision Factors

Three quality factors belonging to this classification are as follows [2, 5, 6]:

Flexibility. Flexibility-related requirements are concerned with the efforts and capabilities required to support adaptive maintenance-related activities. Four elements or subfactors of the flexibility are *generality, self-descriptiveness, modularity,* and *simplicity*.

Testability. Testability-related requirements are concerned with the testing of an information system as well as with its stated operation. Three elements or subfactors of the testability are *user testability, failure maintenance testability,* and *traceability*.

Maintainability. Maintainability-related requirements are concerned with determining the efforts that will be needed by maintenance personnel and users to identify the reasons for the occurrence of software failures, to correct or rectify the failures, and to verify the success of the corrections. Six elements or subfactors of the maintainability are *simplicity, document accessibility, modularity, self-descriptiveness, coding and documentation guidelines,* and *compliance* (consistency).

6.2.2 Product Operation Factors

Five quality factors belonging to this classification are as follows [2, 5, 6]:

Usability. Usability-related requirements are concerned with the scope of the staff resources needed to train a newly hired employee to operate the software system. Two elements or subfactors of the usability are *training* and *operability*.

Efficiency. Efficiency-related requirements are concerned with the hardware resources needed to perform the entire software system functions in conformance with all other requirements. Four elements or subfactors of the efficiency are *communication efficiency, processing efficiency, power usage efficiency* (for portable units), and *storage efficiency*.

Correctness. Correctness-related requirements are outlined in a list of the software system required outputs. Six elements or subfactors of the correctness are *accuracy, coding and documentation guidelines, availability (response time), up-to-dateness, compliance (consistency),* and *completeness*.

Reliability. Reliability-related requirements are concerned with failures to provide a proper level of service. Moreover, they determine the maximum allowed failure rate for the software system and can refer to the entire system or to one or more of its distinct functions. Four elements or subfactors of the reliability are *computational failure recovery, hardware failure recovery, system reliability,* and *application reliability*.

Integrity. Integrity-related requirements are concerned with the security of the software system, i.e., requirements to prevent access to unauthorized people as well as to distinguish between the majority of individuals allowed to view the information ("read permit") and a limited number of individuals who will be allowed to change and add data ("write permit"), etc. Two elements or subfactors of the integrity are *access audit* and *access control*.

6.2.3 Product Transition Factors

Three quality factors belonging to this classification are as follows [2, 5, 6]:

Portability. Portability-related requirements are concerned with the adaptation of a software system under consideration to other environments made up of different hardware, operating systems, etc. Three elements or subfactors of the portability are *software system independence, self-descriptiveness*, and *modularity*.

Interoperability. Interoperability-related requirements are concerned with creating appropriate interfaces with other software systems or with other equipment/product firmware. Four elements or subfactors of the interoperability are *system compatibility, commonality, software system independence*, and *modularity*.

Reusability. Reusability-related requirements are concerned with the use of software modules, originally designed for one specific project, in a new project under development. Seven elements or subfactors of the reusability are *modularity, simplicity, self-descriptiveness, software system independence, document accessibility, application independence*, and *generality*.

6.3 Quality Methods for Use during Software Development

Over the years, many quality methods have been developed to improve software quality during software development. Seven of these methods are shown in Figure 6.1 [7]. Three of the methods shown in Figure 6.1 are described in the following subsections, and detailed information on the remaining four methods is available in the literature [8–10].

6.3.1 Pareto Diagram

This is probably the most effective method in the area of software quality, because past experience clearly indicates that software defects or density

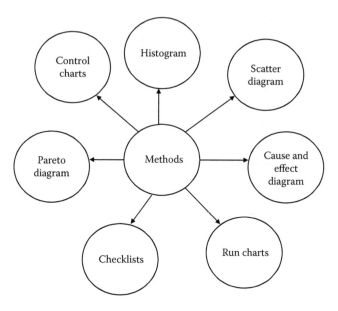

FIGURE 6.1
Some of the quality methods for use during software development.

never follow a uniform distribution. Thus, a Pareto diagram is a very useful tool to highlight focus areas that cause most of the problems in a given software project. For example, Hewlett-Packard has used Pareto diagrams to achieve significant improvements in software quality [11]. Motorola has also successfully utilized Pareto diagrams to highlight the main sources of software-requirement-related changes that enabled in-process corrective actions to be carried out [12]. Additional information on Pareto diagrams in regard to their application during software development is available in the work of Kan [7].

6.3.2 Checklists

Checklists play an important role during software development because they are useful to software programmers/developers in ensuring that all tasks are complete and that the quality characteristics or important factors for each of these tasks are considered. The application of checklists is pervasive. Checklists, used daily by software development personnel, are revised and developed with respect to accumulated experience. Checklists are often a part of the process documents and, as per past experience, their daily application is very useful for keeping the software development processes alive. Additional information on checklists in regard to their application during software development is available in the work of Kan [7].

6.3.3 Run Charts

Run charts are generally utilized for software project management, serving as real-time statements of quality and workload. An example of the application of run charts is tracking the percentage of software fixes that exceed the specified response-time criteria, in order to ensure deliveries of fixes to involved customers in a timely manner. Run charts are also used to monitor the weekly arrival of software defects as well as the defect backlog during the formal testing phases of a machine under consideration. During software development, run charts are frequently compared to the relevant projection models and historical data so that the related interpretations can be placed into proper perspective. Additional information on run charts in regard to their application during software development is available in the work of Kan [7].

6.4 Quality Measures during the Software Development Life Cycle (SDLC)

To produce a quality software product, it is important to take appropriate quality measures during the software development life cycle. A software development life cycle is made up of five stages, as discussed here [13]:

Stage I: Requirements analysis. Past experiences indicate that about 60%–80% of system-development-related failures are due to poor understanding of user requirements [14]. In this regard, during the software development process, major software vendors normally use quality function development (QFD). Software quality function deployment (SQFD) is considered a very useful method to focus on improving the quality of the software development process by implementing appropriate quality improvement approaches to the SDLC requirements solicitation phase. In other words, SQFD is a front-end requirements collection method that quantifiably solicits and defines the customer's critical requirements.

Thus, during SDLC, SQFD is considered a very useful method for solving the problem of poor systems specification. Some of the main benefits of SQFD include fostering better attention to the requirements of customers, establishing better communications among departments and with customers, and quantifying qualitative customer requirements [13].

Stage II: Systems design. This is considered the most critical stage of quality software development because a problem or defect in design is many times more costly to correct than a defect during the production phase or stage. Concurrent engineering is a frequently used method to make changes to systems design, and it is also considered a very useful tool in implementing total quality management [13]. Additional information on this method (i.e., concurrent engineering) is available in the literature [15–17].

Stage III: Systems development. Software total quality management (TQM) requires the proper integration of quality into the total software development process. After the establishment of an effective quality process into Stage I and Stage II of SDLC, the task of coding becomes simple and straightforward [13]. However, for document inspections, the design and code-inspections approach can be used [18]. Furthermore, control charts can be utilized to track the metrics of the effectiveness of code inspections.

Stage IV: Testing. In addition to designing testing activities with care at each stage of the SDLC, such activities must be planned and managed properly right from the start of software development [19]. Furthermore, a TQM-based software development process must have a set of testing objectives. A six-step metric-driven approach can fit quite well with such testing objectives [13]. Its six steps are as follows [2, 13]:

1. Establish structured test objectives.
2. Select appropriate functional methods to derive test-case suites.
3. Run functional tests and assess the degree of structured coverage achieved.
4. Extend the test suites until the desired coverage is achieved.
5. Calculate the test scores.
6. Validate testing by recording errors not found during the testing process.

Stage V: Implementation and maintenance. Most of the software maintenance-related activities are generally reactive. Programmers often zero in on the immediate problem, fix it, and wait until the next problem [13, 20]. As statistical process control (SPC) can be used to monitor the quality of software system maintenance, a TQM-based system must be able to adapt to the SPC process to ensure the quality of maintenance. Additional information on quality during software maintenance is available in the literature [13, 21].

6.5 Software Quality-Related Metrics

Over the years, a large number of metrics have been developed for improving or assuring software quality. Two principal objectives of software quality metrics are (a) to identify conditions that require or enable development of maintenance process improvements in the form of preventive or corrective actions initiated within the organizational structure and (b) to facilitate a proper level of management control, including executing and planning of necessary management interventions. For their successful application, it is absolutely essential that these metrics satisfy the following requirements [2, 6]:

- Easy and simple (i.e., the implementation of the metrics data collection is simple and straightforward and is performed with minimal resources)
- Reliable (i.e., generates similar results when used under similar environments)
- Relevant (i.e., related to an attribute of substantial importance)
- Comprehensive (i.e., applicable to a wide variety of situations and implementations)
- Immune to biased interventions by interested parties
- Valid (i.e., successfully measures the required attribute)
- Mutually exclusive (i.e., does not measure attributes measured by other metrics)
- Does not require independent data collection

A number of useful software quality metrics are discussed in the following subsections [6, 22].

6.5.1 Metric I

This is one of the software process timetable metrics and is defined by

$$TOF = \frac{\theta_1}{\theta_2} \tag{6.1}$$

where
TOF = timetable observance factor
θ_1 = number of milestones completed on time
θ_2 = total number of milestones

6.5.2 Metric II

This is one of the error-severity metrics and is defined by

$$CE_{AS} = \frac{\theta_3}{\theta_4} \tag{6.2}$$

where

CE$_{AS}$ = average severity of code errors
θ_3 = weighted code errors detected
θ_4 = number of code errors detected in the software code through testing and inspections

6.5.3 Metric III

This metric is concerned with measuring the success of help-desk service (HDS) and is expressed by

$$HDS_{SF} = \frac{\theta_5}{\theta_6} \tag{6.3}$$

where

HDS$_{SF}$ = HDS success factor
θ_5 = number of HDS calls completed on time during a 1-year period
θ_6 = total number of HDS calls during a 1-year period

6.5.4 Metric IV

This is one of the HDS productivity metrics and is expressed by

$$HDS_{PF} = \frac{\theta_7}{\theta_8} \tag{6.4}$$

where

HDS$_{PF}$ = HDS productivity factor
θ_7 = total number of yearly working hours invested in help-desk servicing of the software system
θ_8 = the thousands of lines of maintained software code

6.5.5 Metric V

This is one of the error-removal effectiveness metrics and is expressed by

$$DER_E = \frac{\theta_9}{\theta_9 + \theta_{10}} \tag{6.5}$$

where

DER$_E$ = the development error removal effectiveness
θ_{10} = total number of software failures detected during a 1-year period of maintenance service
θ_9 = total number of design and code errors detected in the software development process

Note that data for this metric are normally obtained from design and code reviews and testing reports.

6.5.6 Metric VI

This metric is concerned with measuring the effectiveness of the software corrective maintenance and is defined by

$$CM_E = \frac{\theta_{11}}{\theta_{10}} \tag{6.6}$$

where
CM_E = corrective maintenance effectiveness
θ_{11} = total number of annual working hours invested in corrective maintenance of the software system

6.5.7 Metric VII

This is one of the HDS calls-density metrics and is defined by

$$HDS_{CD} = \frac{\theta_6}{\theta_8} \tag{6.7}$$

where
HDS_{CD} = the HDS calls density

6.5.8 Metric VIII

This is one of the software process productivity metrics and is expressed by

$$SD_P = \frac{\theta_{12}}{\theta_{13}} \tag{6.8}$$

where
SD_P = software development productivity
θ_{12} = number of working hours invested in the software system development
θ_{13} = the thousands of lines of code

6.5.9 Metric IX

This is one of the error-density metrics and is defined by

$$CED = \frac{\theta_4}{\theta_{13}} \tag{6.9}$$

where
CED = code error density

6.5.10 Metrix X

This metric is concerned with measuring the mean severity of the HDS calls and is defined by

$$HDS_{MS\text{-}C} = \frac{\theta_{14}}{\theta_6} \tag{6.10}$$

where

HDS_{MS-C} = the mean severity of HDS calls

θ_{14} = number of weighted HDS calls received during a 1-year period

---

6.6 Software Quality Assurance Manager's Responsibilities and Elements of a Successful Software Quality Assurance Program

There are many responsibilities of a software quality assurance manager. Some of the main ones are as follows [23, 24]:

- Keeping abreast of current software quality matters
- Establishing the software quality control organization, practices, and procedures
- Interfacing with customers
- Disseminating new software information to all concerned groups and individuals
- Keeping management well informed of matters concerning software quality
- Participating in software design-related reviews
- Preparing a software quality program on an annual basis in regard to goals, budgets, manpower, etc.
- Developing new procedures and concepts
- Providing appropriate consulting services to others
- Auditing conformance to the software quality policy on a regular basis
- Training and recruiting personnel familiar with software quality
- Liaising with standardization and regulatory bodies

The following elements or procedures are important to the success of a software quality assurance program [24, 25]:

- Develop a quality assurance activity and ensure its independence.
- Ensure that the quality assurance activity begins at an early stage of the software development cycle.
- Conduct an appropriately detailed verification analysis of the design and requirements.
- Conduct appropriate analysis of the code in addition to testing it.

- Ensure that, prior to the start of the testing process, the development testing is well planned and organized.
- Make sure that documentation is well controlled and cannot be changed without proper controls.
- Keep a careful track of the computer resources required by the end program.
- Carefully evaluate the interfaces between any two elements/parts in the system and resolve any incompatibilities, misunderstandings, and ambiguities.
- Always remain skeptical of errors in software received from any developer.
- Try to conduct some type of quality assurance activity or activities in an environment where you are unable to conduct the ideal maximum of activities.

6.7 Software Quality Assurance Standards and Advantages

Over the years, many software quality assurance-related standards have been developed by various organizations for use in developing software products. Some of the main reasons to have several software quality assurance-related standards are as follows [24]:

- Great diversity in the management approaches, methods, and procedures being used
- Great variety in the types of software products being developed
- Great variety in the technological approaches, methods, and procedures being utilized
- Great diversity in the types of software products being maintained

The main objective of any software quality assurance standard is to produce cost-effective and good-quality products. Thus, top-level management and customers expect that the software quality assurance personnel and management are fully aware of and clearly understand the importance of software quality assurance standards.

Some of the published standards that directly or indirectly concern software quality assurance are as follows [26, 27]:

ISO 9126: Software quality characteristics, prepared by the International Organization for Standardization (ISO). This is an international standard for the evaluation of software and is divided into four parts: quality model, external metrics, internal metrics, and quality in use metrics.

NASA STD 8739.8: Software assurance standard, prepared by the National Aeronautics and Space Administration (NASA). This standard specifies the software assurance-related requirements for software acquired or developed by NASA.

MIL-HDBK-334: Evaluation of contractor's software quality assurance program, prepared by the U.S. Department of Defense. This document is used by the government procurement agency to evaluate the final software quality assurance program in a situation when MIL-S-52779 is applicable to a contract under consideration.

AQAP-14: This document is the North Atlantic Treaty Organization (NATO) equivalent of the preceding document (i.e., MIL-HDBK-334). This document is used to evaluate the software quality control system of NATO contractors.

IEEE-Std-730: IEEE standard for software quality assurance plans, prepared by the Institute of Electrical and Electronics Engineers (IEEE). This standard basically applies to the development and maintenance of critical software.

Additional information on software quality assurance standards is available in the literature [24, 27].

There are many advantages of software quality assurance. Some of the main ones are as follows [28]:

- Reduces project-related risks because of better requirements traceability and thorough testing
- Is useful in enforcing software standards
- Is useful in enhancing the visibility of management into the software development process through reviews and audits
- Ensures that fulfillment of contractual requirements of deliverable items is reviewed by an independent body
- Centralizes the development and maintenance of software approaches
- Centralizes the records related to quality assurance

Problems

1. Write an essay on software quality.
2. Discuss software quality factors.
3. Discuss at least three quality methods that can be used during software development.

4. Compare run charts with a Pareto diagram.

5. Discuss quality measures during the software development life cycle.

6. List at least seven requirements that must be satisfied by software quality metrics for their successful applicability.

7. Define a metric belonging to each of the following three categories:

   a. Software process timetable metrics

   b. Error severity metrics

   c. Software process productivity metrics

8. List at least ten main responsibilities of a software quality assurance manager.

9. List and discuss at least three software quality assurance standards.

10. What are the main advantages of software quality assurance?

References

1. Institute of Electrical and Electronics Engineers (IEEE). (1991). *IEEE standard glossary of software engineering terminology* (IEEE-STD-610-12.1990). New York, NY: Author.
2. Dhillon, B. S. (2007). *Applied reliability and quality: Fundamentals, methods, and procedures*. London: Springer.
3. Mendis, K. S. (1980). A software quality assurance program for the 80s. *Proceedings of the Annual Conference of the American Society for Quality Control*, 379–388.
4. McCall, J., Richards, P., & Walters, G. (1977, November). *Factor in software quality* (NTIS Report No. AD-A049-014, 015, 055). Springfield, VA: National Technical Information Service (NTIS).
5. Evans, M. W., & Marciniak, J. J. (1987). *Software quality assurance and management*. New York, NY: John Wiley and Sons.
6. Galin, D. (2004). *Software quality assurance*. Harlow, Essex, UK: Pearson Education.
7. Kan, S. H. (1995). *Metrics and models in software quality engineering*. Reading, MA: Addison-Wesley.
8. Mears, P. (1995). *Quality improvement tools and techniques*. New York, NY: McGraw-Hill.
9. Kanji, G. K., & Asher, M. (1996). *100 Methods for total quality management*. London: Sage.
10. Ishikawa, K. (1976). *Guide to quality control*. Tokyo: Asian Productivity Organization.
11. Grady, R. B., & Caswell, D. L. (1986). *Software metrics: Establishing a company-wide program*. Englewood Cliffs, NJ: Prentice Hall.

12. Daskalantonakis, M. K. (1992). A practical view of software measurement and implementation experiences with Motorola. *IEEE Transactions on Software Engineering, SE-18*, 998–1010.
13. Gong, B., Yen, D. C., & Chou, D. C. (1998). A manager's guide to total quality software design. *Industrial Management and Data Systems, 98*(3), 100–107.
14. Gupta, Y. P. (1988). Directions of structured approaches in system development. *Industrial Management and Data Analysis, 88*(7), 11–18.
15. Dhillon, B. S. (2002). *Engineering and technology management: Tools and applications.* Boston, MA: Artech House.
16. Salomone, T. A. (1995). *Concurrent engineering.* New York, NY: Marcel Dekker.
17. Rosenblatt, A., & Watson, G. F. (1991). Concurrent engineering. *IEEE Spectrum, 28*(7), 22–23.
18. Fagan, M. E. (1986). Advances in software inspection. *IEEE Transactions on Software Engineering, 12*(7), 744–751.
19. Graham, D. R. (1992). Testing and quality assurance: The future. *Information and Software Technology, 34*(10), 694–697.
20. Osborne, W. M. (1988). All about software maintenance: 50 questions and answers. *Journal of Information Systems Management, 5*(3), 36–43.
21. Schulmeyer, G. G., & McManus, J. I. (1992). *Total quality management for software.* New York, NY: Van Nostrand Reinhold.
22. Schulmeyer, G. G. (1999). Software quality assurance metrics. In G. G. Schulmeyer & J. I. McManus (Eds.), *The handbook of quality assurance* (pp. 403–443). Upper Saddle River, NJ: Prentice Hall.
23. Tice, G. D. (1983). Management policy and practices for quality software. *Proceedings of the Annual American Society for Quality Control Conference*, 369–372.
24. Dhillon, B. S. (1987). *Reliability in computer system design.* Norwood, NJ: Ablex Publishing.
25. Rubey, R. J. (1977). Planning for software reliability. *Proceedings of the Annual Reliability and Maintainability Symposium*, 495–499.
26. Dunn, R., & Ullman, R. (1982). *Quality assurance for computer software.* New York, NY: McGraw-Hill.
27. Fisher, M. J. (1981). Software quality assurance standards: The coming revolution. *Journal of Systems and Software, 2*, 357–362.
28. Fischer, K. F. (1978). A program for software quality assurance. *Proceedings of the Annual Conference of the American Society for Quality Control*, 333–340.

7

Human Error and Software Bugs in Computer Systems

7.1 Introduction

Human error and software bugs have become a pressing issue in computer systems because of problems such as high cost, catastrophic failures, and incorrect decisions and actions. For example, a study commissioned by the U.S. Department of Commerce's National Institute of Standards and Technology (NIST) in 2002 reported that software bugs or errors alone cost the U.S. economy about $59 billion annually (i.e., around 0.6% of its gross domestic product (GDP) [1]. Furthermore, in regard to catastrophic failures, in 1996 the European Space Agency's US$1-billion prototype Ariane 5 rocket was destroyed just forty seconds after launch due to a bug in the onboard guidance computer program [2].

Finally, in regard to incorrect decisions and actions, in 1997 the Smart Ship USS *Yorktown* was left dead in the water for almost three hours after a divide-by-zero error [3]. Human error and software bugs are pressing issues in computer systems, and this chapter presents some important aspects of human error and software bugs in computer systems.

7.2 Facts, Figures, and Examples

Some of the facts, figures, and examples concerned with human error and software bugs in computer systems are as follows:

- Up to the end of 1992, the number of people killed due to computer system failures worldwide was between 1,000 and 3,000 [4, 5].
- A software bug in the code controlling the Therac-25 radiation therapy machine caused many deaths in the 1980s [6].

- In 1989, a pilot set the heading in a plane's computer control inertial navigation system as 270° instead of 027°; the plane ran out of fuel, resulting in twelve fatalities [7].
- In 1963, a software error caused the incapacitation of a North American Air Defense Command (NORAD) exercise [8].
- In 2002, a study commissioned by the National Institute of Standards and Technology reported that software bugs or errors cost the U.S. economy around $59 billion per year [1].
- In 1991, because of a software error, a MIM-104 Patriot (surface-to-air missile system) failed to intercept an incoming Iraqi Scud missile that resulted in twenty-eight American fatalities in Saudi Arabia [9].
- In 1981, the launching of the first U.S. space shuttle was postponed for about twenty minutes prior to the scheduled launching time because of a software fault [10].
- In 1996, the US$1-billion prototype Ariane 5 rocket of the European Space Agency was destroyed just forty seconds after launch due to a bug in the onboard guidance computer program [2].
- A computer opened the vent valve on the wrong vessel because of a software fault, and fourteen tons of carbon dioxide were vented and lost [7, 11].

7.3 Factors Affecting the Occurrence of Human Error in Computer Systems

There are many factors that affect the occurrence of human error in computer systems. They may be categorized under seven classifications, as shown in Figure 7.1 [12]. These classifications, along with their approximate percentage contribution to human error (in parentheses), are: personal (35%), design factors (20%), human/computer interface (10%), written instructions (10%), training (10%), organizational accuracy requirements (10%), and environmental (5%) [12].

- The classification *personal* includes those factors that are concerned with traits, conditions, or characteristics peculiar to humans that affect, in a relatively consistent manner, their ability to carry out assigned tasks properly. The classification includes items such as skills and knowledge, physiological and psychological needs, sensory processes, sleep loss, drug use, fatigue, motivation, and body rhythms.

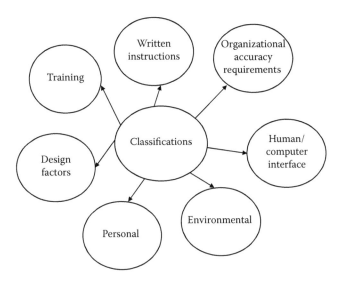

FIGURE 7.1
Classifications of factors affecting the occurrence of human error in computer systems.

- The classification *design factors* refers to any overall system design condition that may lower human reliability. The classification includes items such as function allocation (i.e., distribution of tasks between computer and humans), sufficient time (i.e., time allowed to complete manual activities), and error-related feedback to involved humans.
- The classification *human/computer interface* includes those factors concerned with such items as input devices, messages, command language, and data codes.
- The classification *written instructions* includes those factors concerned with such items as performance aids (i.e., documents storing ready-to-use information by system operators while performing an activity) and preparation of instructions.
- The classification *training* includes those factors concerned with various aspects of training, including types of training, learning principles, computer-based training, and special training needs.
- The classification *organizational accuracy requirements* includes those factors concerned with such items as setting of accuracy requirements and error-related reporting mechanisms.
- The classification *environmental* includes those factors concerned with the physical environment (e.g., noise and room temperature) and the social environment.

Additional information on all seven of these classifications of factors is available in the work by Bailey [12].

7.4 Computer Failure Categories and Hardware and Software Error Sources

Computer failures may be classified into the following five categories [7]:

Hardware failures. These are just like failures in any other piece of equipment, and they occur due to factors such as poor design, defective parts, poor maintenance, and unexpected environmental conditions.

Software failures. These are the result of the inability of programs to continue processing due to erroneous logic.

Human errors. These are the result of inappropriate actions or lack of actions by humans involved in the process (e.g., the system's operators, designers, and builders).

Specifications failures. These are distinguished by their origins, i.e., defects in the specification of the system, rather than in the execution or design of either software or hardware.

Malicious failures. These are due to a relatively new phenomenon, i.e., the malicious introduction of programs intended to cause damage to anonymous users. Often these programs are referred to as *computer viruses*.

There are many sources of hardware and software errors. Some of these are as follows [12]:

- Inherited errors
- Data preparation errors
- Handwriting errors
- Keying errors
- Optical character reader

Inherited errors can account for over 50% of the errors in a computer-based system [12]. Furthermore, data-preparation tasks can also generate a significant proportion of errors. As per Bailey [12], at least 40% of all errors come from manipulating the data (i.e., data preparation) prior to writing it down or entering it into the computer system. Additional information on computer failure categories and hardware and software error sources is available in the literature [7, 12].

7.5 Common Software Errors in Programming

There are many commonly occurring software errors in programming. These may be categorized under the following eight classifications [13–15]:

Logic errors. Examples include an excessive or inadequate number of statements in a loop, logical expressions containing incorrect operands, and improper sequencing of logic activities.

Data definition errors. Examples include ill-defined data and data referred out of bounds.

Computational errors. Examples include missing computations, equations containing a wrong operand, and wrongly used parentheses.

Interface errors. Examples include calling a subroutine at a wrong place or not making the call at all.

Data handling errors. Examples include incorrect data initialization and variables referred to by incorrect names.

Database errors. Examples include incorrectly expressed data units, data not properly initialized in the database, and initialization of data to a wrong value.

Input/output errors. Examples include data not written at all, input read from an incorrect data file, and incorrectly written data.

Miscellaneous errors. These errors do not fall into any of the other seven classifications.

7.6 Factors Causing Human Errors during Software Development and Maintenance

There are many factors that can cause human errors during software development and maintenance. These may be grouped under two classifications, as discussed in the following subsections [13, 16, 17].

7.6.1 Classification I Factors

Two major factors belonging to this classification are *software perceivability* and *rigor of software definition*. Software perceivability may simply be described as a measure of the psychometric complexity of software components. More specifically, perceivability is a measure of complexity as experienced by individuals attempting to comprehend a system with an

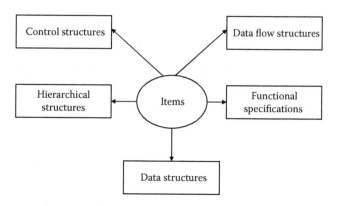

FIGURE 7.2
Items with a significant effect on software reliability that require unambiguous and complete definitions.

intention to perform tasks concerned with that system. Two examples of such tasks are modifying the system and examining it with respect to faultlessness. The capability of individuals to comprehend an abstract system depends upon the following system-complexity aspects:

- Syntactic complexity
- Pragmatic complexity
- Semantic complexity

The *syntactic complexity* aspect is related to system topology or, more specifically, to the number of constituent parts or components and the nature and variety of part or component interconnections. The *pragmatic complexity* aspect is related to the system presentation format. Finally, the *semantic complexity* aspect is related to the nature of functions performed by the system components.

The other major factor, *rigor of software definition*, is a measure of unambiguity and the completeness of the definition of the software component. Note that software reliability is significantly affected by the items shown in Figure 7.2, which require definitions that are unambiguous and complete [15].

7.6.2 Classification II Factors

These factors are usually referred to as the *project factors* and are concerned with the characteristics of software engineering project components. These factors are within the control of the project manager. During the software development and maintenance phase, there are various factors that cause human errors. These may be grouped under the following four categories.

1. **Environmental factors.** Environmental factors include noise level, heating and ventilation levels, intensity of light, level of pressure to accomplish job within a specified time limit, team values, team morale, and management attitudes toward reliability.
2. **Tools (for programmers) factors.** The availability of effective tools may affect the occurrence of human error and the detection and removal of faults caused by human error. During software development or maintenance, programmers use tools such as code generators, methodologies, standards, computer aids for designing and testing, rules and guidelines, and programming languages.
3. **Personal factors.** These factors are associated with personnel in regard to their skills, knowledge, and attitudes.
4. **Task-related factors.** These factors are concerned with the characteristics of software engineering tasks.

7.7 Methods To Prevent Programmers from Inadvertently Introducing Bugs during the Software Writing Process

There are many methods to prevent programmers from inadvertently introducing bugs during the software writing process. Some of these methods are as follows [18, 19]:

Programming style. Various innovations in programming style and defensive programming are designed to make the occurrence of software bugs less likely or easier to spot.

Programming language support. Programming languages include features such as static type systems, modular programming, and restricted name spaces to help programmers prevent bugs.

Development methodologies. There are several schemes/methodologies to manage programmer activity so that fewer bugs are generated. Many of these schemes/methodologies fall under the discipline of software engineering (which looks into software design-related issues as well). For example, formal program specifications are employed to state the exact behavior of all involved programs so that design-related bugs can be eliminated.

Code analysis. Tools for code analysis help software developers by inspecting the program text beyond the capability of the compiler to highlight potential difficulties or problems. Although the problem of discovering all types of programming-related errors in a given specification is generally not solvable, these tools exploit the fact that programmers tend to make the same kinds of mistakes during the software writing process.

Additional information on these methods is available in the literature [18, 19].

7.8 Software Error–Related Metrics

Many metrics have been developed to measure the extent of errors, faults, and failures in the system artifacts during the life-cycle phases. Some useful metrics concerned with software error are presented in the following subsections [5, 20, 21].

7.8.1 Metric I: Defect Density

This metric calculates the ratio of defects per lines of code or lines of design and is expressed by

$$D_d = \left(\sum_{j=1}^{n} D_j \right) \Big/ A \text{ (or } B) \tag{7.1}$$

where

D_d = defect density
n = number of inspections (or life-cycle phases) to date
D_j = total number of unique defects discovered during the jth inspection process of a life-cycle phase
A = total number of executable source code statements plus data declarations during the implementation phase and beyond
B = total number of source lines of design statements (in thousands) during the design phase

As per IEEE [20], this metric (i.e., Metric I) can be used during all eight life-cycle phases. IEEE [20] also gives the metric an experience code of 3, indicating that it has received a high level of operational validity in the industrial sector. Additional information on the metric is available in the literature [5, 20, 21].

7.8.2 Metric II: Cumulative Failure Profile

This metric simply provides a graphical depiction of the cumulative number of failures discovered during the life-cycle phases, such that [5, 20, 21]

TF_j = total number of failures discovered during life-cycle phase j

The metric is quite flexible, i.e., the unit for which the failures are counted can be the complete system, by module, or by subsystem. Furthermore, the interval for which the failures are counted can be subphases or whole phases. As per IEEE [20], the metric can be used during all life-cycle phases. IEEE [20] also gives the metric an experience code of 1, indicating that it has received a low level of operational validity by the industrial sector. Additional information on this metric is available in the literature [5, 20, 21].

7.8.3 Metric III: Fault Density

This metric calculates the ratio of faults per line of code and is expressed by [5, 20, 21]:

$$FD = NF/\alpha \qquad (7.2)$$

where
  FD = fault density
  NF = number of unique faults discovered
  α = lines of executable source code (in thousands), including data
      declarations

As per IEEE [20], the metric can be used during all life-cycle phases. IEEE [20] also gives the metric an experience code of 2, indicating that it has received a reasonable level of operational validity in the industrial sector.

To enhance usefulness of this metric for mission-critical systems, two variations to the metric can be made [5]. First, in order to minimize the problems encountered with lines of code (LOC), an equivalent calculation can be carried out by substituting function points (FPs) for α. Second, if the fault's severity is distinctive, the calculation can be enlarged or expanded to determine fault density by severity. Additional information on this metric is available in the work of Herrman [5].

7.8.4 Metric IV: Defect Indices

This metric calculates a relative index of correctness of software throughout the different life-cycle phases and is expressed by [5, 20, 21]

$$I_d = \sum jX_j/N_j \qquad (7.3)$$

where
  I_d = defect index
  N_j = product size at jth phase, which can be measured in A, B, or FPs

$$X_j = [\theta_1 \lambda_j / \mu_j] + [\theta_2 \beta_j / \mu_j] + [\theta_3 \gamma_j / \mu_j] \tag{7.4}$$

where

  θ_1 = weighting factor for serious defects (default is 10)
  θ_3 = weighting factor for trivial defects (default is 1)
  θ_2 = weighting factor for medium defects (default is 3)
  λ_j = number of serious defects discovered during the *j*th phase
  γ_j = number of trivial defects discovered during the *j*th phase
  β_j = number of medium defects discovered during the *j*th phase
  μ_j = total number of defects detected during the *j*th phase

Note that the value of X_j is calculated at the end of each of eight phases and is weighted by phase, such that j = 1,2,3,…,8. As per IEEE [20], the metric from Equation (7.3) can be applied in all life-cycle phases. IEEE [20] also gives the metric an experience code of 1, indicating that it has received a low level of operational validity in the industrial sector.

  This metric can be modified for application in mission-critical systems by inserting the four International Electrotechnical Commission (IEC) standard severity levels for the three provided and then making adjustment to the weighting factors as follows [5]:

$$I_d = \sum j X_j / N_j \tag{7.5}$$

where

$$X_j = [\theta_1 d_{1j} / \mu_j] + [\theta_2 d_{2j} / \mu_j] + [\theta_3 d_{3j} / \mu_j] + [\theta_4 d_{4j} / \mu_j] \tag{7.6}$$

where

  θ_1 = weighting factor for catastrophic defects (default is 10)
  θ_2 = weighting factor for critical defects (default is 8)
  θ_3 = weighting factor for marginal defects (default is 3)
  θ_4 = weighting factor for negligible defects (default is 1)
  d_{1j} = number of catastrophic defects discovered during the *j*th phase
  d_{2j} = number of critical defects discovered during the *j*th phase
  d_{3j} = number of marginal defects discovered during the *j*th phase
  d_{4j} = number of negligible defects discovered during the *j*th phase
  N_j = product size at *j*th phase

Additional information on this metric is available in the literature [5, 20, 21].

Problems

1. List the five most important facts and figures concerned with human error and software bugs in computer systems.
2. Write an essay on human error and software bugs in computer systems.

3. What are the factors that affect the occurrence of human error in computer systems?

4. List and discuss the main categories of computer failures.

5. What are the common software errors in programming?

6. Discuss the factors that cause human error during software development and maintenance.

7. Discuss at least four methods concerned with preventing programmers from inadvertently introducing bugs during the software writing process.

8. Define the following two software error related metrics:

   a. Defect density

   b. Fault density

9. Compare logic errors with computational errors.

10. Compare hardware failures with software failures.

References

1. National Institute of Standards and Technology (NIST). (2002). Gaithersburg, MD: Author.

2. Dowson, M. (1997). The Ariane 5 software failure. *Software Engineering Notes, 22*(2), 84–85.

3. U.S. Navy. (1997). *USS* Yorktown. Washington, DC: Department of Defense.

4. Kletz, T. (1997). Reflections on safety. *Safety Systems, 6*(3), 1–3.

5. Herrman, D. S. (1999). *Software safety and reliability.* Los Alamitos, CA: IEEE Computer Society Press.

6. Leveson, N., & Turner, C. S. (1993). An investigation of the Therac-25 accidents. *IEEE Computer, 26*(7), 18–41.

7. Kletz, T., Chung, P., Broomfield, E., & Shen-Orr, C. (1995). *Computer control and human error.* Houston, TX: Gulf Publishing.

8. Myers, G. J. (1976). *Software reliability principles and practices.* New York, NY: John Wiley and Sons.

9. U.S. Government Accounting Office (GAO). (1992). *Patriot missile defense: Software problem led to system failure at Dhahran, Saudi Arabia* (Report No. IMTEC 92-26). Washington, DC: Author.

10. Graman, J. R. (1981). The bug heard around the world. *Software Engineering Notes, 6*(October), 3–10.

11. Nimmo, I. (1994). Extend HAZOP to computer control systems. *Chemical Engineering Progress, 90*(10), 32–44.

12. Bailey, R. W. (1983). *Human error in computer systems.* Englewood Cliffs, NJ: Prentice Hall.

13. Dhillon, B. S. (1987). *Reliability in computer system design.* Norwood, NJ: Ablex Publishing.

14. Thayer, T. A., Lipow, M., & Nelson, E. C. (1978). *Software reliability: A study of large project reality.* Amsterdam: North-Holland Publishing.
15. Joshi, R. D. (1983). Software development for reliable software systems. *Journal of Systems and Software, 3,* 107–121.
16. Rzevski, G. (1982). Identification of factors which cause software failure. *Proceedings of the Annual Reliability and Maintainability Symposium,* 157–161.
17. Lauber, R. J. (1982). Impact of a computer-aided development support system on software quality and reliability. *Proceedings of the Annual IEEE Computer Society International Conference* (COMPSAC), 248–255.
18. Huizinga, D., & Kolawa, A. (2007). *Automated defect prevention: Best practices in software management.* New York, NY: Wiley-IEEE Computer Society Press.
19. McDonald, M., Musson, R., & Smith, R. (2007). *The practical guide to defect prevention.* Seattle, WA: Microsoft Press.
20. IEEE. (1989). *IEEE standard dictionary of measures to produce reliable software* (ANSI/IEEE STD 982.1-1989). New York, NY: Author.
21. IEEE. (1989). *IEEE guide for the use of IEEE standard dictionary of measures to produce reliable software* (ANSI/IEEE STD 982.2-1989). New York, NY: Author.

8

Software Safety and Internet Reliability

8.1 Introduction

Each year billions of dollars are spent to develop various types of software around the globe, and safety has become an important issue. More specifically, in many applications, proper functioning is so crucial that a simple malfunction may lead to a large-scale loss of lives and a high cost. For example, commuter trains in Paris, France, serve about 800,000 passengers daily and depend on software signaling [1].

The history of the Internet may be traced back to the late 1960s with the development of the Advanced Research Projects Agency Network (ARPANET) [2]. It has grown from 4 hosts in 1969 to around 147 million hosts and 38 million sites in 2002, and today over 2.1 billion people used the services of the Internet around the globe [2, 3]. In 2001, there were over 52,000 Internet-related failures and incidents. The reliability and stability of the Internet has become extremely important to the global economy and other areas, because Internet failures can cause millions of dollars in losses and interrupt the day-to-day routines of millions of end users [4]. This chapter presents various important aspects of software safety and Internet reliability.

8.2 Software Safety Classifications and Potential Hazards

Software safety may be categorized under three classifications, as shown in Figure 8.1 [5]. These are safety-critical software, software-related software, and non-safety-related software. The safety-critical software controls or performs such functions, and if they are executed erroneously or if they failed to execute properly, they could directly inflict serious injuries to humans and/or the environment and cause deaths. Similarly, the safety-related software controls or performs such functions which are activated to minimize or prevent altogether the effect of a safety-critical system failure. Finally,

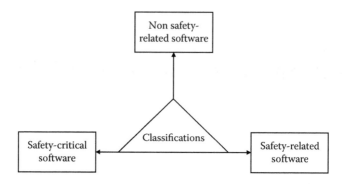

FIGURE 8.1
Software safety classifications.

the non-safety-related software controls or performs those system functions that are not concerned with safety. All in all, most mission-critical systems incorporate a combination of safety-related, non-safety-related, and safety-critical software [5].

There are many ways in which software can cause/contribute to a hazard [6–8]:

- Performed a function not required
- Failed to perform a required function
- Provided wrong solution to a problem
- Performed a function out of sequence
- Failed to recognize a hazardous situation requiring a corrective action
- Poor response to a contingency

8.3 Software Risk Classifications

There are basically three ways in which software can increase risk [9]:

1. Failure to detect and take an appropriate corrective measure to recover from a hazardous situation
2. Failure to mitigate the damage after the occurrence of an accident
3. Directing the system toward a hazardous direction or state

Software risks, in which losses can occur, may be grouped under the following three classifications [10]:

Classification I. This classification includes situations where environmental conditions may impact the ability of a software professional to carry out his/her job effectively. Some examples of such conditions are poor lighting, screen glare, inadequate computer memory/hardware, and the wrong software development tools for the job.

Classification II. This classification includes situations such as no compliance to policies and procedures, inadequate standards, and no usage of standards. An example of such situations is writing a software code prior to properly completing the requirements definition.

Classification III. This classification includes situations such as poorly trained workers to carry out assigned tasks, unavailability of required manuals, and poorly planned or funded projects.

8.4 Basic Software System Safety-Related Tasks

There are many software system safety-related tasks. Some of the basic ones are as follows [8, 11]:

- Highlight safety-critical variables and elements for use by code developers.
- Trace identified system-associated hazards to the software-hardware interface.
- Develop system-specific software design requirements and criteria, computer–human interface-related requirements, and testing requirements on the basis of highlighted software system safety-related constraints.
- Develop safety-related software test case requirements, test descriptions, test procedures, and test plans.
- Trace safety-associated constraints and requirements right up to the code level.
- Identify software components that, directly or indirectly, control safety-critical operations and then direct safety analysis and tests on those particular functions as well as on the safety-critical path that leads to their execution.
- Show the software system safety-constraint-related consistency in regard to the software requirements specification.
- Perform any special safety analysis, e.g., computer–human interface analysis or software fault tree analysis.
- Review the test results concerning safety issues and trace the identified safety-related software shortcomings back to the system level.

- Develop an effective tracking system within the software along with system configuration control structure to ensure traceability of safety requirements and their flow through documentation.

8.5 Software Safety Assurance Program and Software Quality Assurance Organization's Role in Regard to Software Safety

A software safety assurance program within an organizational setup basically involves the following three maturity levels [8, 12]:

Maturity level A. This is concerned with the development of an organization's/company's culture that recognizes the importance of safety-related issues. More specifically, company software developers perform their tasks according to standard development rules and apply them quite consistently.

Maturity level B. This is concerned with the implementation of a development process that involves safety assurance reviews as well as hazard analysis for identifying and eliminating safety-critical situations prior to being designed into the system.

Maturity level C. This is concerned with the utilization of a design process that documents results as well as implements continuous-improvement methods to eliminate safety-critical errors in the system software.

Some of the items that need to be considered with care during the implementation of a software safety assurance program are as follows [8]:

- All human–computer interface requirements and software system safety requirements are consistent with contract requirements.
- All software system-related hazards are identified, evaluated, tracked, and eliminated as per requirements.
- Changes in mission requirements, configuration, or design are carried out such that they clearly maintain an acceptable level of risk.
- Software system safety is clearly quantifiable to the specified level of risk using the general measuring methods.
- Software system safety is addressed with respect to team effort that involves groups such as quality assurance, management, and engineering.
- Software system safety-associated requirements are specified and developed as an element of the organization's design policy.

- Past software safety data are considered with care and used in all future software development projects.

A software safety assurance program must also consider factors such as presented here [8, 12]:

- Record all types of safety-related data.
- Ensure that safety is designed into the system cost effectively and in a timely fashion.
- Minimize the retrofit actions required to improve safety by including appropriate safety features during research and development in an effective manner.
- Carry out necessary changes in design configuration and user requirements such that they maintain an acceptable risk level.
- Minimize risk when using and accepting new designs, materials, production and test methods, etc.

A software quality assurance organization plays various roles in regard to software safety. Some of these roles are as follows [8, 12]:

- Define user-safety-related requirements, the operational doctrine, and the operational concept.
- Establish the operational safety policy that clearly highlights acceptable risks and operational alternatives to hazardous operations.
- Define appropriate requirements to perform operational safety reviews.
- Evaluate, investigate, resolve, and document all reported safety-related operational incidents.
- Conduct safety audits and reviews of operational systems on a regular basis.
- Determine appropriate safety-related criteria for system acceptance.
- Approve the findings of safety testing prior to releasing the systems.
- Chair operational safety-related review panels.

8.6 Software Hazard Analysis Methods

There are a large number of methods that can be used to perform various types of software hazard analysis. Some of these methods are as follows [8, 13–16]:

- Code walk-through
- Software fault tree analysis

- Software sneak circuit analysis
- Proof of correctness
- Failure modes and effect analysis
- Event-tree analysis
- Petri net analysis
- Hazard and operability studies
- Cause-consequence diagrams
- Software/hardware integrated critical path
- Nuclear safety cross-check analysis

The first five of these methods are described in the following subsections. Additional information on the remaining methods is available in the literature [13–16].

8.6.1 Code Walk-Through

This is a quite useful method to improve software safety and basically is a team effort among involved professionals such as software programmers, program managers, the system safety people, and software engineers. Code walk-throughs entail a rigorous review of the software through inspection and discussion of the software functionality. All logic branches as well as the function of each statement are discussed thoroughly.

The system reviews the functionality of software and compares it with the system-associated requirements. This verifies that all software safety-related requirements are implemented appropriately, in addition to determining the accuracy of functionality. Additional information on the method is available in the work by Sheriff [16].

8.6.2 Software Fault Tree Analysis (SFTA)

SFTA is used to analyze the safety of a software design and is an offshoot of Fault Tree Analysis (FTA) described in Chapter 3 [17]. The two main objectives of SFTA are as follows [18]:

- Demonstrate that the logic contained in the software design will not generate system safety-related failures.
- Determine environmental conditions that may result in the software causing a safety failure.

SFTA proceeds in a similar manner to hardware fault tree analysis (i.e., FTA). More specifically, FTA begins by identifying an undesirable event known as the *top event* of a system under consideration. Fault events that could cause the occurrence of the top event are generated and connected by logic operators normally referred to as AND and OR gates. The AND gate generates an output if all its input fault events occur. In contrast, the OR gate produces an output if one or more of its input fault events occur.

The construction of a fault tree of a system/situation under consideration proceeds by generation of fault events successively until the fault events need not be developed any further. A fault tree may simply be described as a logic structure that relates the top event to basic or primary events.

Although fault trees for hardware and software are developed separately, at their interfaces they are linked together to allow total system analysis. This is very important, since it is impossible to develop software safety procedures in isolation; rather, they must be considered as a component of the entire system safety. For example, a software error may lead to a mishap only if a hardware failure occurs simultaneously.

Finally, note that although SFTA is an effective hazard analysis method, it is costly to use. Additional information on FTA is available in the work of Dhillon and Singh [17].

8.6.3 Software Sneak Circuit Analysis

This method is used to identify software logic that leads to undesirable outputs. Program source code is converted to topological network trees, and six basic patterns are employed to model the code: entry dome, return dome, iteration loop, parallel line, single line, and trap. Each and every software mode is modeled using these basic patterns linked in a network tree flowing right from top to bottom. The involved individual (i.e., analyst) asks questions concerning the interrelationships and use of the instructions that are considered the structural elements. The answers to the questions asked provide clues that highlight sneak conditions that may lead to undesirable outputs.

The analysts search for the following four basic software sneaks:

- Presence of an undesired output
- The undesired inhibit of an output
- Incorrect timing
- A program message poorly describing the actual condition

The clue-generating questions are taken from the topograph representing the code segment, and whenever sneaks are discovered, the analysts conduct investigative analyses to verify that the code indeed generated the sneaks. Subsequently, the impacts of the sneaks are assessed and necessary corrective actions recommended. Additional information on software sneak circuit analysis is available in the work of Ericson [19].

8.6.4 Proof of Correctness

Proof of correctness is another method used to conduct software hazard analysis, and it decomposes a program into a number of logical segments. Input/output assertions for each of these segments are defined. Subsequently, a software professional verifies from the perspective that each input assertion and its output assertion are true and that if all input assertions are true, then all output assertions are also true. Finally, note that the method makes use of mathematical theorem-proving concepts to verify that a program under consideration is consistent with its specifications. Additional information on this method is available in the work of Weik [20].

8.6.5 Failure Modes and Effect Analysis (FMEA)

This method—developed in the early 1950s to perform failure analysis of flight control systems [21, 22]—is sometimes used to perform software hazard analysis. Basically, the method demands the listing of all possible failure modes of each part/element/component and their possible effects on the listed subsystems, system, etc. This method is described in detail in Chapter 3 and in the work of Dhillon [22].

8.7 Software Standards and Useful Software Safety Design-Related Guidelines

There are various types of standards for use in developing software. They are useful to provide reference points to indicate minimum acceptable requirements or to compare systems [23]. Some of the standards directly or indirectly concerned with software safety are as follows [8, 12, 23]:

IEEE 1228-1994, Software Safety Plans, Institute of Electrical and Electronic Engineers (IEEE), New York, 1994.

IEC 60880, Software for Computers in the Safety Systems of Nuclear Power Stations, International Electrotechnical Commission (IEC), Geneva, Switzerland, 1986.

NSS 1740.13 Interim, Software Safety Standard, National Aeronautics and Space Administration (NASA), Washington, DC, 1994.

MIL-STD-882D, System Safety Program Requirements, Department of Defense, Washington, DC, 2000.

EIA SEB6A, System Safety Engineering in Software Development, Electronic Industries Alliance (EIA), New York, 1990.

ANSI/AAMI, Risk Management—Part I: Application of Risk Management, American National Standards Institute (ANSI), New York, 2000.

IEEE 1059-1993, Guide for Software Verification and Validation Plans, Institute for Electrical and Electronic Engineers (IEEE), New York, 1993.

DEF STD 00-55-1, Requirements for Safety-Related Software in Defence Equipment, Ministry of Defence, London, 1997.

There are many useful software safety design-related guidelines developed over the years by professionals working in the field of computers. A careful application of such guidelines can be very useful to improve software safety. Some of these guidelines are as follows [8, 24]:

- Develop appropriate software modules to monitor critical software in regard to errors, faults, timing problems, or hazardous states.
- Ensure that all conditional statements meet all possible conditions effectively and are under full software control.
- Remove obsolete or unnecessary code.
- Avoid using all 0 or 1 digits for critical variables.
- Develop software design such that it effectively prevents inadvertent/ unauthorized access and/or any modification to the code.
- Do not allow safety-critical software patches throughout the development process.
- Incorporate necessary provisions to detect and log system errors.
- Initialize spare memory with a bit pattern that, if ever accessed and executed, will clearly direct the involved software toward a safe state.
- Include the requirement for a password along with confirmation before the execution of a safety-critical software module.
- Separate and isolate safety-critical software modules from non-safety-critical software modules.
- Include appropriate mechanisms to ensure that safety-critical computer software parts and interfaces are under positive control at all times.
- Include an operator to authorize or validate the execution of safety-critical commands.

8.8 Internet Facts, Figures, and Failure Examples; Benefits and Impediments of Internet Electronic Commerce; and Internet Reliability-Related Observations

Some Internet facts, figures, and failure examples are as follows:

- In 2011, over 2.1 billion people around the globe were using the Internet, and about 45% of them were below the age of twenty-five years [25].

- In 2000, Internet-related economy generated about $830 billion in revenues in the United States [2, 4].
- In 2000, the Internet carried 51% of the information flowing through two-way telecommunication, and by 2007 over 97% of all telecommunicated information was carried over the Internet [26].
- From 2006 to 2011, developing countries increased their share of the world's total number of Internet users from 44% to 62% [25].
- In 2001, there were 52,658 Internet-related failures and incidents [2, 4].
- On April 25, 1997, a misconfigured router of a Virginia service provider injected an incorrect map into the global Internet and, in turn, the Internet providers who accepted this map automatically diverted their traffic to the Virginia provider [27]. This caused network congestion, instability, and overload of Internet router table memory that ultimately shut down many of the main Internet backbones for about two hours [2, 27].
- On November 8, 1998, a malformed routing control message due to a software fault triggered an interoperability problem between a number of core Internet backbone routers produced by different vendors. In turn, this caused a widespread loss of network connectivity in addition to an increment in packet loss and latency [2]. It took a number of hours for most of the backbone providers to overcome this outage.
- On August 14, 1998, a misconfigured main Internet database server incorrectly referred all queries for Internet machines/systems with names ending in "net" to the wrong secondary database server. In turn, because of this problem, the majority of connections to "net" Internet web servers and other end stations malfunctioned for a number of hours [2].

Interest in conducting electronic commerce (EC) on the Internet continues to increase at a significant rate. A study of executives' opinions in regard to conducting business on the Internet reported many benefits and impediments [28]. The benefits included ease of access and global reach, low-cost advertising medium, low barriers to entry, and perceived image enhancement. In contrast, the impediments were centered on six major topics: security, legal issues, startup costs, uncertainty and lack of information, lack of skilled personnel, and training and maintenance. Additional information on these benefits and impediments is available in the work of Nath, Akmanligil, Hjelm, Sakaguchi, and Schultz [28].

A study by Lapovitz, Ahuja, and Jahamian [29] reported the following Internet reliability-related observations:

- Mean time to failure and mean time to repair for most of the Internet backbone paths are around twenty-five days or less and twenty minutes or less, respectively.

- The Internet backbone structure's mean time to failure and availability are significantly lower than the Public Switched Telephone Network (PSTN).
- Most interprovider path malfunctions result from congestion collapse.
- In the Internet backbone infrastructure, there is only a minute fraction of network paths that contribute disproportionately to the number of long-term outages and backbone unavailability.

8.9 Classifications of Internet Outages

A case study of Internet outages conducted over a period of one year has categorized the outages under twelve classifications (along with their occurrence percentages in parentheses), as shown in Figure 8.2 [29].

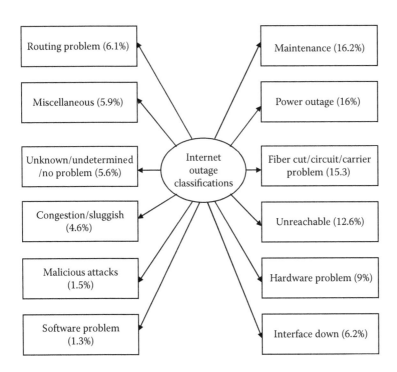

FIGURE 8.2
Classifications of Internet outages (along with their occurrence percentages in parentheses).

8.10 An Approach for Automating Fault Detection in Internet Services and Models for Performing Internet Reliability and Availability Analysis

Many Internet services (e.g., e-commerce and search engines) suffer faults, and a quick detection of these faults could be an important factor in improving system availability. For this very purpose, an approach called the *pinpoint method* is considered very useful. This method combines the low-level monitors' easy deploy-ability with the higher-level monitors' ability to detect application-level faults [30]. The method is based upon the following three assumptions in regard to the system under observation and its workload [30]:

- There is a considerably higher number of basically independent requests (i.e., from different users).
- An interaction with the system is relatively short-lived, the processing of which can be decomposed as a path or, more specifically, a tree of the names of parts/elements that participate in the servicing of that request.
- The software is composed of a number of interconnected modules with clearly defined narrow interfaces, which could be software subsystems, objects, or simply physical mode boundaries.

The pinpoint method is a three-stage process: observing the system, learning the patterns in system behavior, and detecting anomalies in system behaviors [30]. The stage "observing the system" is concerned with capturing the runtime path of each and every request handled or served by the system and then, from these paths, extracting two particular low-level behaviors that are likely to reflect high-level functionality (i.e., interactions of components/ parts and path shapes).

The stage "learning the patterns in system behavior" is concerned with constructing a reference model that clearly represents the usual behavior of an application with respect to component/part interactions and path shapes. The model is constructed under the assumption that most of the system functions normally most of the time.

Finally, the stage "detecting anomalies in system behaviors" is basically concerned with analyzing the ongoing behaviors of the system and detecting anomalies in regard to the reference model. Additional information on this method is available in the work of Kiciman and Fox [30].

There are many mathematical models that can be used to perform various types of reliability and availability analysis concerning the reliability of Internet services [2, 22, 31–34]. Two of these models are presented in the following subsections.

8.10.1 Model I

This model is concerned with evaluating the availability of an Internetworking (router) system made up of two identical and independent switches. The model assumes that the switches form a standby-type configuration and that the system fails when both of the switches fail. In addition, the switch failure and restoration rates are constant. The system state space diagram is shown in Figure 8.3, and the numerals in boxes and circles denote system states.

The following symbols were used to develop equations for the model:

j = the jth system state shown in Figure 8.3 for: $j = 0$ (system operating normally [i.e., two switches functional: one operating, the other on standby]), $j = 1$ (one switch operating, the other failed), $j = 2$ (system failed [both switches failed])

p = the probability of failure detection and successful switchover from switch failure

λ = the switch constant failure rate

μ_1 = the constant repair/restoration rate from system state 2 to state 0

μ = the switch constant restoration/repair rate

$P_j(t)$ = the probability that the Internetworking (router) system is in state j at time t, for $j = 0,1,2$

With the aid of the Markov method, we write down the following differential equations for the diagram in Figure 8.3 [22, 35]:

$$\frac{dP_0(t)}{dt} + [p\lambda + (1-p)\lambda]P_0(t) = \mu P_1(t) + \mu_1 P_2(t) \qquad (8.1)$$

$$\frac{dP_1(t)}{dt} + (\lambda + \mu)P_1(t) = p\lambda P_0(t) \qquad (8.2)$$

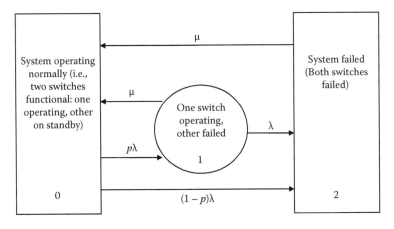

FIGURE 8.3
System state space diagram.

$$\frac{dP_2(t)}{dt} + \mu_1 P_2(t) = \lambda P_1(t) + (1-p)\lambda P_0(t) \tag{8.3}$$

At time $t = 0$, $P_0(0) = 1$ and $P_1(0) = P_2(0) = 0$.

The following steady-state probability solutions are obtained by setting derivatives equal to zero in Equations (8.1)–(8.3) and using the relationship $\sum_{j=0}^{2} P_j = 1$.

$$P_0 = \mu_1(\mu + \lambda)/B \tag{8.4}$$

where

$$B_0 = \mu_1(\mu + p\lambda + \lambda) + (1-p)\lambda(\mu + \lambda) + p\lambda^2 \tag{8.5}$$

$$P_1 = p\lambda\mu_1/B \tag{8.6}$$

$$P_2 = [p\lambda^2 + (1-p)\lambda(\mu + \lambda)]/B \tag{8.7}$$

where
   $P_j =$ the steady-state probability that the Internetworking (router) system is in state j, for $j = 0,1,2$.

The system steady-state availability is given by

$$AV_{ss} = P_0 + P_1$$

$$= [\mu_1(\mu + \lambda) + p\lambda\mu_1]/B \tag{8.8}$$

where
   $AV_{ss} =$ the system steady-state availability.

8.10.2 Model II

This mathematical model is concerned with evaluating the reliability and availability of a server system. The model assumes that the Internet server system can either be in an operating or a failed state. In addition, its (i.e., Internet server system) failure/outage and repair/restoration rates are constant, and all its failures/outages occur independently and the restored/repaired server system is as good as new.

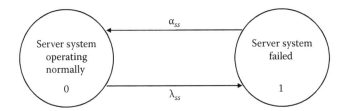

FIGURE 8.4
Server system state space diagram.

The system state space diagram is shown in Figure 8.4, and the numerals in circles denote system states. The following symbols were used to develop equations for the model:

j = the jth system state shown in Figure 8.4 for: $j = 0$ (server system operating normally), $j = 1$ (server system failed)

λ_{ss} = the server system constant outage/failure rate

α_{ss} = the server system constant restoration/repair rate

$P_j(t)$ = the probability that the server system is in state j at time t, for $j = 0,1$

With the aid of Markov method, we write down the following differential equations for the diagram in Figure 8.4 [22]:

$$\frac{dP_0(t)}{dt} + \lambda_{ss} P_0(t) = \alpha_{ss} P_1(t) \tag{8.9}$$

$$\frac{dP_1(t)}{dt} + \alpha_{ss} P_1(t) = \lambda_{ss} P_0(t) \tag{8.10}$$

at time $t = 0$, $P_0(0) = 1$, and $P_1(0) = 0$.

By solving Equations (8.9)–(8.10), we get the following state probability equations:

$$P_0(t) = AV_{ss}(t) = \frac{\alpha_{ss}}{(\lambda_{ss} + \alpha_{ss})} + \frac{\lambda_{ss}}{(\lambda_{ss} + \alpha_{ss})} e^{-(\lambda_{ss} + \alpha_{ss})t} \tag{8.11}$$

$$P_1(t) = UA_{ss}(t) = \frac{\lambda_{ss}}{(\lambda_{ss} + \alpha_{ss})} - \frac{\lambda_{ss}}{(\lambda_{ss} + \alpha_{ss})} e^{-(\lambda_{ss} + \alpha_{ss})t} \tag{8.12}$$

where

$AV_{ss}(t)$ = the server system availability at time t

$UA_{ss}(t)$ = the server system unavailability at time t

As time t becomes very large, Equations (8.11) and (8.12) reduce to

$$AV_{ss} = \lim_{t \to \infty} AV_{ss}(t) = \frac{\alpha_{ss}}{\lambda_{ss} + \alpha_{ss}} \tag{8.13}$$

$$UA_{ss} = \lim_{t \to \infty} UA_{ss}(t) = \frac{\lambda_{ss}}{\lambda_{ss} + \alpha_{ss}} \tag{8.14}$$

where
AV_{ss} = the server system steady-state availability
UA_{ss} = the server system steady state unavailability

For $\alpha_{ss} = 0$, Equation (8.11) reduces to

$$R_{ss}(t) = e^{-\lambda_{ss}t} \tag{8.15}$$

where
$R_{ss}(t)$ = the server system reliability at time t

Thus, the server system mean time to failure is given by [22]

$$MTTF_{ss} = \int_0^\infty R_{ss}(t)\,dt$$

$$= \int_0^\infty e^{-\lambda_{ss}t}\,dt \tag{8.16}$$

$$= \frac{1}{\lambda_{ss}}$$

where
$MTTF_{ss}$ = the server system mean time to failure

Example 8.1

Assume that the constant failure and repair rates of an Internet server system are 0.005 failures/hour and 0.08 repairs/hour, respectively. Calculate the server system availability for a 10-hour mission.
By inserting the specified data values into Equation (8.11), we obtain

$$AV_{ss}(10) = \frac{0.08}{(0.005 + 0.08)} + \frac{0.005}{(0.005 + 0.08)} e^{-(0.005+0.08)(10)}$$

$$= 0.9663$$

Thus, the server system availability for the specified mission time is 0.9663.

Problems

1. Write an essay on software safety and Internet reliability.
2. What are the ways in which software can cause/contribute to a hazard?
3. Discuss basic software system safety-related tasks.
4. What are the important items that should be considered with care during the implementation of a software safety assurance program?
5. List at least ten methods that can be used to perform software hazard analysis.
6. Describe software sneak circuit analysis.
7. What are the useful software safety design-related guidelines?
8. Discuss at least three examples of Internet failures.
9. What are the benefits and impediments of Internet electronic commerce?
10. Prove Equations (8.4)–(8.7) by using Equations (8.1)–(8.3).

References

1. Cha, S. S. (1993). Management aspect of software safety. *Proceedings of the Eighth Annual Conference on Computer Assurance*, 35–40.
2. Dhillon, B. S. (2007). *Applied reliability and quality: Fundamentals, methods, and procedures*. London: Springer.
3. Hafner, K., & Lyon, M. (1996). *Where wizards stay up late: The origins of the Internet*. New York, NY: Simon and Schuster.
4. Goseva-Popstojanova, K., Mazidar, S., & Singh, A. D. (2004). Empirical study of session-based workload and reliability for web servers. *Proceedings of the 15th International Symposium on Software Reliability Engineering*, 403–414.
5. Herrmann, D. S. (1999). *Software safety and reliability*. Los Alamitos, CA: IEEE Computer Society Press.
6. Leveson, N. G. (1986). Software safety: Why, what, and how? *Computing Surveys, 18*(2), 125–163.
7. Friedman, M. A., & Voas, J. M. (1995). *Software assessment*. New York, NY: John Wiley and Sons.
8. Dhillon, B. S. (2003). *Engineering safety: Fundamentals, techniques, and applications*. River Edge, NJ: World Scientific Publishing.
9. Leveson, N. G., Cha, S. S., & Shimeall, T. J. (1991). Safety verification of ADA programs using software fault trees. *IEEE Software, 8*(4), 48–59.
10. Fortier, S. C., & Michael, J. B. (1993). A risk-based approach to cost-benefit analysis of software safety activities. *Proceedings of the Eighth Annual Conference on Computer Assurance*, 53–60.

11. Leveson, N. G. (1995). *Safeware*. Reading, MA: Addison-Wesley Publishing.
12. Mendis, K. S. (1999). Software safety and its relation to software quality assurance. In G. G. Schulmeyer & J. I. McManus (Eds.), *Handbook of software quality assurance* (pp. 669–679). Upper Saddle River, NJ: Prentice Hall.
13. Ippolito, L. M., & Wallace, D. R. (1998, January). *A study on hazard analysis in high integrity software standards and guidelines* (Report No. NISTIR 5589). Washington, DC: National Institute of Standards and Technology, U.S. Department of Commerce.
14. Hammer, W., & Price, D. (2001). *Occupational safety management and engineering*. Upper Saddle River, NJ: Prentice Hall.
15. Hansen, M. D. (1989). Survey of available software-safety analysis techniques. *Proceedings of the Annual Reliability and Maintainability Symposium*, 46–49.
16. Sheriff, Y. S. (1992). Software safety analysis: The characteristics of efficient technical walk-throughs. *Microelectronics and Reliability, 32*(3), 407–414.
17. Dhillon, B. S., & Singh, C. (1981). *Engineering reliability: New techniques and applications*. New York, NY: John Wiley and Sons.
18. Leveson, N. G., & Harvey, P. R. (1983). Analyzing software safety. *IEEE Transactions on Software Engineering, 9*(5), 569–579.
19. Ericson, C. A. (2005). *Hazard analysis techniques for system safety*. New York, NY: John Wiley and Sons.
20. Weik, M. H. (2000). *Computer science and communications dictionary* (Vol. 2). Norwell, MA: Kluwer Academic Publishers.
21. Institute of Electrical and Electronic Engineers (IEEE). (1994, May). *Software safety plans* (IEEE 1228-1994). New York, NY: Author.
22. Dhillon, B. S. (1999). *Design reliability: Fundamentals and applications*. Boca Raton, FL: CRC Press.
23. Wallace, D. R., Kohn, D. R., & Ippolito, L. M. (1992). An analysis of selected software safety standards. *Proceedings of the Seventh Annual Conference on Computer Assurance*, 123–136.
24. Keene, S. J. (1992). Assuring software safety. *Proceedings of the Annual Reliability and Maintainability Symposium*, 274–279.
25. International Telecommunication Union. (2011). *ICT facts and figures*. Geneva, Switzerland: ICT Data and Statistics Division, Telecommunication Development Bureau.
26. Hilbert, M., & Lopez, P. (2011, April). The world's technological capacity to store, communicate, and compute information. *Science, 332*(6025), 60–65.
27. Barrett, R., Haar, S., & Whitestone, R. (1997, April 25). Routing snafu causes Internet outage. *Interactive Week*, p. 9.
28. Nath, R., Akmanligil, M., Hjelm, K., Sakaguchi, T., & Schultz, M. (1998). Electronic commerce and the Internet: Issues, problems, and perspectives. *International Journal of Information Management, 18*(2), 91–101.
29. Lapovitz, C., Ahuja, A., & Jahamian, F. (1999). Experimental study of Internet stability and wide-area backbone failures. *Proceedings of the 29th Annual International Symposium on Fault-Tolerant Computing*, 278–285.
30. Kiciman, E., & Fox, A. (2005). Detecting application-level failures in component-based Internet services. *IEEE Transactions on Neural Networks, 16*(5), 1027–1041.
31. Hecht, M. (2001). Reliability/availability modeling and prediction of e-commerce and other Internet information systems. *Proceedings of the Annual Reliability and Maintainability Symposium*, 176–182.

32. Aida, M., & Abe, T. (2001). Stochastic model of Internet access patterns. *IEICE Transactions on Communications, E 84-B*(8), 2142–2150.

33. Chan, C. K., & Tortorella, M. (2001). Spare-inventory sizing for end-to-end service availability. *Proceedings of the Annual Reliability and Maintainability Symposium*, 98–102.

34. Imaizumi, M., Kimura, M., & Yasui, K. (2005). Optimal monitoring policy for server system with illegal access. *Proceedings of the 11th ISSAT International Conference on Reliability and Quality in Design*, 155–159.

35. Dhillon, B. S., & Kirmizi, F. (2003). Probabilistic safety analysis of maintainable systems. *Journal of Quality in Maintenance Engineering, 9*(3), 303–320.

9

Software Usability

9.1 Introduction

Over the years, with the increasing development of software for interactive use, attention to the requirements and preferences of potential end users has intensified quite significantly. Nowadays, the user interface often plays a crucial role in the success or failure of a software project. Furthermore, according to Myers and Robson [1], about 50%–80% of all source code development accounts, directly or indirectly, for the user interface.

In general, user-friendly software enables its potential users to perform their tasks easily and intuitively, and it clearly supports rapid learning and high skills retention of the involved individuals. Furthermore, in today's competitive global environment, usability is not a luxury, but a fundamental ingredient in software systems, as the users' comfort and productivity relate directly to it.

Software usability may be defined as quality in software application or use [2]: specifically, how productively its users will be able to perform their tasks, how much support the users will need, and how easy and straightforward the software is to learn and use [3]. This chapter presents various important aspects of software usability extracted from the published literature.

9.2 Need for Considering Usability during the Software Development Phase and Basic Principles of the Human-Computer Interface

To produce user-friendly software products, careful consideration to usability during the development phase is essential. Some of the important factors that dictate the need to consider usability during the software development process are as follows [4]:

Competition. Failure to address usability-related issues properly can lead to the loss of market share, should competitors release their software products with better usability.

Cost. Poor usability software products reduce productivity of user and increase developer cost with respect to customer support service, hotlines, etc.

Mixed users. Users of the software products could be professionals or nonprofessionals with limited or no computer skills at all.

Global market. Software products normally cover a global market with varying language proficiencies, cultures, etc.

Five basic principles of the human–computer interface in regard to software are as follows [4, 5]:

Principle I. The software system must be able to provide an effective feedback to potential users/workers in regard to their performance.

Principle II. The software must meet all types of task-related requirements.

Principle III. The software system must be able to display information in a format as well as at a pace adapted for potential operators.

Principle IV. The software must be easy to use, user friendly, and adaptable to the levels of knowledge or experience of potential operators.

Principle V. The software-related ergonomics principles must be applied properly, in particular, to human data processing.

9.3 Software Usability Engineering Process

Past experience indicates that the software usability engineering process may be viewed differently by different organizations. Nonetheless, a typical process followed in product development is essentially composed of three principal activities in parallel, as shown in Figure 9.1 [6].

Activity I is concerned with gaining insight into customers' current experience with a system. Data on users' experience are obtained basically through contextual interviews (i.e., relevant interviews are conducted while users perform their tasks). During the interviews, users are asked about their system interfaces, the type of work being performed, perception of the system, etc. One important benefit of the contextual interviews is that they produce large amounts of data quite rapidly. However, it is to be noted with care that different users in different contexts have different needs. The aspects of a user's context that influence the usability of a system for each user include items such as the type of work being performed,

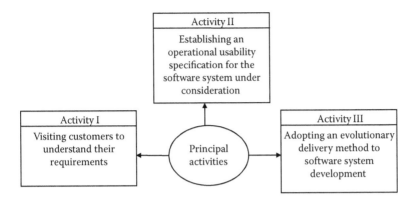

FIGURE 9.1
Three principal activities followed in parallel in the software product development process.

interactions with other software systems, organizational culture, and physical workplace environment.

In the case of Activity II, a software usability specification may be described as a measurable definition of usability that is shared by all of the people involved. This is based on the understanding of the user needs, the resources needed to produce the software system, and competitive analysis. Two important points to be considered with care in developing the usability specification are as follows [6]:

- People involved with development of the usability specification must evaluate it continuously during the development process and make necessary changes to reflect up-to-date information on the needs of users.
- Failure to understand the user needs prior to developing a specification can result in a specification document that does not reflect the needs of users.

Activity III means first building a small subset of the software system and then "growing" it during the development process and studying users on a continuous basis as the system evolves. Furthermore, it is noteworthy that evolutionary delivery exploits, rather than overlooks, the dynamic nature of software-related requirements. Some of the methods used to improve software system usability during evolutionary delivery stages include collecting user feedback during early field tests, building and testing early prototypes, and performing analysis of the impact of design solutions [7–9]. An important advantage of the evolutionary delivery method is that it helps to build the shared understanding of project team members of the user-interface design of the software system.

9.4 Steps for Improving Software Product Usability

Software usability can be improved by following the four steps presented below [4, 10].

Step I: Understand users. Two particular items essential to understanding users are the software product usage environment and the user profile. The software product use environment is useful to understand the range of factors that impact the utilization of the product. Some of the environmental factors taken into consideration include the type of network and security, system configuration, browser types and settings (if applicable), degree of privacy/noise levels, location (office, mobile, home, etc.), and connection speed (if applicable).

The user profile is useful to focus design efforts on actual issues concerning potential users and avoid wasting resources and time on sideline issues. However, the user profile requires the collection of information on items such as potential users' interests, needs, and demographics. Some of the sources to obtain this type of information are market research data, product registrations, marketing and sales staff, and training and customer support staff.

Step II: Evaluate software product under consideration. Usability is relative. It simply means that a software product may be simple and straightforward to use for one type of users, but quite confusing or intuitive for other users. Thus, an evaluation from the perspective of users seeks to determine if there is compatibility between the software product and its potential users. A good method of evaluating a software product includes actions such as testing usability, determining the user-product "fit," analyzing the available user data, and performing user field research (i.e., ethnographic research).

The action *testing usability* is basically concerned with determining where users are having difficulties in using the software product. Usability testing can be performed in different settings, including on customer premises, a designated area within the framework of company establishment, and an outside market research facility. The action *determining the user-product "fit"* is concerned with assessing and prioritizing all usability-related issues. This should be carried out by keeping in mind that the most critical interface elements that, directly or indirectly, impact the success of a software product include presentation performance, screen language clarity, graphic quality, intuitive navigation, personalization of content, error handling, and underlying behavioral metaphor.

The action *analyzing the available user data* is concerned with analyzing the data obtained from sources such as customer service, technical support, marketing people, and sales support. Finally, the action *performing user field research* is concerned with obtaining first-hand information on the functioning of the software product with its intended users.

Step III: Assess available resources. This step is basically concerned with assessing the available resources and the capability of all involved personnel to execute the software product design/redesign project. The diverse capabilities and talents of all the design team members should also be taken into consideration with care. Furthermore, it is extremely important to assess the existence of user advocacy among the individuals forming the design team, because this is key to producing an effective user-centered design. There should always be at least one design team member to advocate for the users' point of view.

Step IV: Update development process. In order to have an effective usability process for future uses/releases, an evaluation of the overall development cycle is absolutely necessary. The critical elements of a user-centered design process are keeping satisfactory design-related documentation, establishing cycles of user feedback, and establishing approaches of tracking results.

Nonetheless, it should be noted with care that the documents that are useful for software design/redesign work include user-interface specification, flowcharts, functional specifications, application specifications, and marketing (business) requirements. Finally, it is worth emphasizing that the evaluation of documents such as these is a very important step toward updating the development process.

9.5 Software Usability Inspection Methods and Considerations for Their Selection

There are many software usability inspection methods. Five widely used methods to inspect/evaluate software usability are shown in Figure 9.2 [4, 10, 11]. Each of these methods is described in the following list:

Cognitive walkthrough. This method uses a detailed procedure to simulate task execution at each step of the dialogue to determine, if the simulated user's memory content and goals can be safely

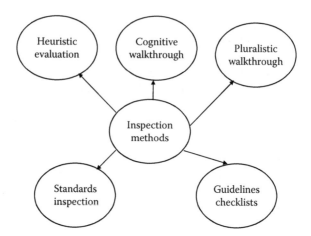

FIGURE 9.2
Common software usability inspection methods.

assumed to lead to the next correct anticipated measure or action. The principal advantages of this method are that it is an effective approach to predicting problems and capturing the cognitive process. In contrast, its two main disadvantages are that it is focused on one attribute of usability and that there is a need to train a skilled evaluator [12].

Heuristic evaluation. This method involves usability specialists who determine whether each dialogue element meets set usability principles effectively. Some of the main advantages of this method are straightforward to learn and use, useful to highlight problems early in the design process, and inexpensive to implement.

Pluralistic walk-through. This method makes use of group meetings where individuals such as usability developers, specialists, and students step through a learning scenario and discuss all dialogue elements. Furthermore, this feature inspection lists items such as sequences of features used to carry out typical tasks, steps that need extensive experience to properly evaluate a proposed set of features, checks for long sequences, and difficult steps. Some of the main advantages of this method are that it is straightforward to learn and use, permits iterative testing/evaluation, and is useful in satisfying the criteria of all involved parties.

Standards inspection. This method involves usability experts who inspect the interface for compliance with given standards. These standards could be domain-specific software standards, departmental standards (if any), or user-interface standards.

Guidelines checklists. These are useful to ensure that the appropriate usability-related principles will be considered in software-related design work. A checklist provides inspectors with a basis for comparing the software product. Usually, checklists are employed in conjunction with a usability evaluation/inspection method.

Additional information on these five methods is available in the literature [10, 11].

During the selection of an appropriate usability inspection/evaluation method or combination of methods for a specific application, the individual involved in the selection process must take into consideration the different foci of the evaluation. Most of these foci or considerations are as follows [13]:

- The resources required
- The type of measures provided
- Evaluation style
- The information provided
- The stage in the life cycle at which the evaluation is performed
- The level of interference implied
- The method's subjectivity/objectivity level
- The immediacy of the response

9.6 Software Usability Testing Methods and Important Factors in Regard to Such Methods

There are a number of usability testing methods that measure system performance with respect to predefined criteria according to the usability attributes specified by the usability standards and empirical metrics [5]. Normally, in these methods or approaches, users carry out certain tasks with the software product/system. Furthermore, the required data are collected on measured performance (e.g., the time needed to perform the task).

Four commonly used software usability testing methods are as follows [5]:

Method I: Thinking-aloud protocol. This is a widely used method in the area of software usability testing because it is helpful in conducting formative evaluations [14]. During the test, participants are asked to express their feelings, opinions, and thoughts while interacting with the software product and carrying out tasks. Their remarks provide significant insight into the most effective method of designing the system interaction. Note, however, that the thinking-aloud protocol could be quite difficult to use with certain user groups (e.g., young students), who are distracted by the process.

Method II: Performance measurement. This is another commonly used method in which usability tests are directed at determining quantitative and hard data. Generally, these data are in the form of performance metrics (e.g., required time to carry out specific tasks). Finally, it is worth noting that the International Organization for Standardization (ISO) promotes the usability evaluation method based on measured performance of predetermined usability metrics [15].

Method III: In-field studies. This method is basically concerned with observing the users performing their assigned tasks in their normal work/study environment. The principal advantages of the method are the natural user performance and group interaction. However, the method has certain limitations in terms of measuring performance because appropriate testing equipment cannot be used properly in normal work environments.

Method IV: Codiscovery. In this method, a group of users carry out tasks together that simulate a work process under observation. Furthermore, most of these users have someone else available for help. The codiscovery method is clearly considered useful in various work scenarios.

Finally, factors such as those shown in Figure 9.3 are very important in regard to software usability testing methods [16]. Additional information on these four methods is available in the literature [5, 13, 16].

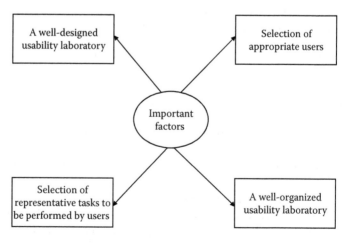

FIGURE 9.3
Important factors with respect to software usability testing methods.

9.7 Guidelines for Conducting Software Usability Testing

Over the years, professionals working in the area of usability have developed many useful guidelines to perform software usability testing. Seven of these guidelines are presented here [4, 12].

Guideline I: Treat each person participating in the test as a completely new case. This means that each participant should be considered unique irrespective of his/her previous performance in usability testing sessions or background.

Guideline II: Keep the testing session as neutral as possible. This means that there should be no vested interest, whatsoever, in the outcomes of a given test.

Guideline III: Use the "thinking aloud" method as considered appropriate. This method has proved to be very useful in capturing the thinking of the participating people while working with the interactive software.

Guideline IV: Ensure that the individual directing/performing the usability test is clearly conscious of his/her body language and voice. This is important because it is fairly easy to unintentionally influence others through body language and voice.

Guideline V: Assist people taking part in the usability test only in exceptional circumstances. More clearly, let these individuals struggle as much as possible.

Guideline VI: Ensure that the person directing/performing the usability test keeps going even when he/she makes an error. This individual must not panic even when he/she has inadvertently revealed some information or in some other way has clearly biased the usability test in process. If the involved individual just continues, then his/her action may not even be observed by the people taking part in the test.

Guideline VII: Use humor as necessary to keep the test environment fairly relaxed. Humor has proved to be quite useful in counteracting the self-consciousness of participants, and it can also help them to relax.

Problems

1. Write an essay on software usability.
2. Discuss the important factors that dictate the need to consider usability during the software development phase.
3. What are the basic principles of the human–computer interface with respect to software?

4. Describe the software usability engineering process.

5. Discuss the steps to improve software product usability.

6. What are the commonly used methods to inspect/evaluate software usability?

7. List at least seven factors that must be considered during the selection of software usability inspection methods.

8. Describe the following three software usability testing methods:

   a. Thinking-aloud protocol

   b. Performance measurement

   c. In-field studies

9. Discuss at least five useful guidelines to conduct software usability testing.

10. What are the important factors with respect to software usability testing methods?

References

1. Myers, B., & Robson, M. B. (1992). Survey on user interface programming. *Proceedings of the ACM CHI'92 Human Factors in Computing Systems Conference*, 195–202.

2. International Organization for Standardization (ISO). (1999). *Software product evaluation: General overview* (ISO/IEC 14598-1). Geneva, Switzerland: Author.

3. Juristo, N., Windl, H., & Constantine, L. (2001). Introducing usability. *IEEE Software, 18*(January/February), 20–21.

4. Dhillon, B. S. (2004). *Engineering usability: Fundamentals, applications, human factors, and human error.* Stevenson Ranch, CA: American Scientific Publishers.

5. Avouris, N. M. (2002). *An introduction to software usability.* ECE Department, University of Patras, Rio Patras, Greece.

6. Good, M. (1988). Software usability engineering. *Digital Technical Journal, 6*(February), 125–133.

7. Wixon, D., & Bramhall, M. (1985). How operating systems are used: A comparison of VMS and UNIX. *Proceedings of the Human Factors Society 29th Annual Meeting*, 245–249.

8. Whiteside, J., Archer, N., Wixon, D., & Good, M. (1982). How do people really use text editors? *SIGOA Newsletter, 3*(1–2), 29–40.

9. Good, J., Whiteside, D., Wixon, D., & Jones, S. (1984). Building a user-derived interface. *Communications of the ACM, 27*(October), 1032–1043.

10. Emerson, M., Porter, D., & Rudman, F. (2002). *Improving software usability: A manager's guide.* East Chatham, NY: Enervision Media.

11. Fitzpatrick, R. (2002). *Strategies for evaluating software usability.* Department of Mathematics, Statistics, and Computer Science, Dublin Institute of Technology, Dublin, Ireland.

12. Lee, S. H. (1999). Usability testing for developing effective interactive multimedia software: Concepts, dimensions, and procedures. *Educational Technology and Society*, 2(2), 100–113.

13. Dix, A., Finlay, J., Abowd, G., & Beale, R. (1998). *Human computer interaction*. Hemel Hempstead, UK: Prentice Hall.

14. Ferre, X., Juristo, N., Windl, H., & Constantine, L. (2001). Usability basics for software developers. *IEEE Software*, 18(1), 22–29.

15. International Organization for Standardization (ISO). (1997). *Ergonomics requirements for office work with visual display terminals (VDT), Part II: Guidance on usability* (ISO 9241-11, draft international standard). Geneva, Switzerland: Author.

16. Preece, J., Rogers, Y., Sharp, H., Benyon, D., Holland, S., & Carey, T. (1994). *Human computer interaction*. Reading, MA: Addison-Wesley.

10

Web Usability

10.1 Introduction

The World Wide Web (WWW) was released a couple of decades ago by the European Laboratory for Particle Physics (CERN), and its growth has mushroomed to hundreds of millions of sites around the globe from a mere 623 sites in 1993 [1]. Today, the usage of the web in the world economy has become an instrumental factor. For example, in 2001, the global e-commerce market was estimated to be about $1.2 trillion, and today it has grown to many trillions of dollars [1, 2]. Moreover, there are billions of web users around the globe.

Web usability may simply be expressed as allowing the users to manipulate features of a website to accomplish certain goals [1, 2]. Some of the main goals of web usability can be described as follows [2, 3]:

- Present the information to potential users in a clear and concise fashion.
- Remove any ambiguity whatsoever concerning the consequences of an action (e.g., clicking on delete/remove/purchase).
- Put the most important thing in the proper place on a web page or a web application.
- Provide the correct choices to the potential users, and do so in a very obvious way.

Today, usability rules the web, because if a website is not easy and straightforward to use, people simply leave and move on to something else. This chapter presents various important aspects of web usability.

10.2 Web Usability Facts and Figures

Some relevant facts and figures related to web usability are as follows:

- In 2000, over 50% of the companies in the United States sold their products online, and there were over 800 million pages on the web throughout the United States [4, 5].
- A study revealed that about 70% of retailers lacked a well-defined e-commerce strategy and firmly believed that they were using their websites to test the waters for online-related demand [6].
- A study of e-commerce sites reported that only about 56% of intended tasks were carried out successfully by the site users [7].
- User interface accounts for roughly 47%–60% of the lines of system or application code [8].
- A research study reported that about 65% of all online shopping trips result in failure [9].
- Studies have shown that about 10% of all users scroll beyond the information that is visible on the screen when a web page appears [2, 10, 11].
- A study revealed that around 40% of website users elected not to return to a site because of design problems [12].
- Studies have shown that web usability is improving by roughly 2%–8% per year [2, 11, 13].

10.3 Common Web Design Errors

There are many errors that are common on all levels of web design. Some of these are as follows [11]:

Page design error. This often occurs when the emphasis is on creating attractive pages to evoke good feelings about the company/organization. Instead, the emphasis should be on designing for an optimal user experience under a day-to-day environment. Utility is more important than attractive pages.

Linking strategy error. This occurs when the designer treats the website as an indispensable entity and thus does not provide proper links to other sites. Beyond not having appropriate entry points where others can link, many organizations even overlook the use of essential links to their own site in their very own advertisements.

Content authorizing error. This occurs when the designer writes in the normal linear style instead of writing for potential online readers who frequently scan text and need short pages, where secondary information is best relegated to supporting pages.

Information architecture–related error. This occurs when the website is constructed to mirror the organizational structure rather than structuring it to mirror the tasks of users and their specific views of the information space.

Business model–related error. This occurs when the designer treats the web as a marketing communication (Marcom) brochure rather than recognizing it as a paradigm shift that will ultimately change the way that business-related transactions are carried out in this age of a networked economy.

Project management–related error. This occurs when a web project is managed simply as a conventional corporate project rather than as a single-customer-interface project. The principal drawback of the traditional method is that it generally leads to a rather internally focused design with an inconsistent user interface.

10.4 Web Page Design

The design of a web page is a crucial factor in regard to the effectiveness of web usability, as it is the most immediately visible element of web design. Some of the important usability dos and don'ts in regard to page design are presented in Table 10.1 [14].

TABLE 10.1

Web Page Design: Important Usability Dos and Don'ts

No.	Dos	Don'ts
1	Consider screen real estate as a highly valuable commodity.	Avoid using all capital letters.
2	Keep the size of most web pages to a level that can be downloaded within 10 seconds.	Avoid specifying fonts using absolute sizes.
3	Design web pages such that the potential browsers can easily resize them to meet their specific needs.	Avoid assuming that users can see what you see.
4	Tailor images to elements that are clearly meaningful. Furthermore, past experiences indicate that dense graphics generally alienate users.	Avoid using animation unless it is essential.
5	Fit main page contents in the potential browser window's width, even when the window is not maximized to fill the total screen.	Avoid getting carried away with creative or "artistic" fonts.
6	Make use of visual highlighting as necessary to draw attention of users to important information.	...

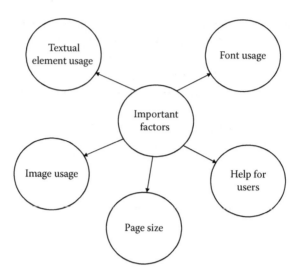

FIGURE 10.1
Some important factors to be considered in web page design.

Some of the important factors that should be considered with care in page design are shown in Figure 10.1 [14]. Each of the important factors shown in Figure 10.1 is described in the following subsections.

10.4.1 Image Usage

Users often blame webpage images as an impediment to successful web access. In this regard, the following guidelines to the use of images may be helpful [2, 11, 14]:

- Reduce the image resolution as much as possible.
- Limit the use of different colors.
- Use animation only when it clearly adds to the meaning of the information.
- Reuse graphics on other pages if the need arises.
- Use a commercial image-compression tool to reduce the size of image files as much as possible.
- Use the most efficient format for an image.
- Use a thumbnail image on web pages that link to the larger image.
- Limit graphics to those elements that are really needed.
- Avoid using a different photograph on each and every page of a website, as this will degrade the website's performance.

10.4.2 Page Size

Page size is important to usability in two ways: (1) the downloading and displaying speed of pages and (2) the flexibility of the pages to fit the available display area. In regard to "the downloading and displaying speed of pages," the length of time for downloading a page from the server and displaying it in the browser window is an important factor to sizing web pages effectively. Nonetheless, response time may simply be defined as the time from when a user requests a page to when it has been displayed totally. Some of the useful guidelines concerned with response time are as follows:

- Ensure that the response time is within one second to fit into the chain of thoughts of potential users.
- Provide an adequate warning to potential users when a web page will need more than ten seconds to download.
- Ensure that the response time is within 0.1 second in order to make the system feel interactive.
- Ensure that the response time is well within ten seconds to keep the attention of potential users.

In regard to "the flexibility of the pages to fit the available display area," some of the useful guidelines for its successful achievement are as follows:

- During the design process, pay close attention to the ability to resize headers, footers, etc.
- Design web pages in such a way that they can easily be resized (i.e., to fit within a wide range of window sizes).
- Ensure that the key page elements are clearly visible with scrolling when the window is 400 pixels in height.
- Use relative, instead of absolute, sizes for elements that fall under a browser's resizing capability.
- Generally, no horizontal scrolling is required when the window is 800 pixels wide.

10.4.3 Textual Element Usage

Most of the key elements of a website are conveyed through the use of text, tables, and lists. Thus it is very important to write in a style that not only transmits appropriate information, but also reflects how websites are used. Some of the guidelines considered useful in writing effective web page text are presented here [14].

- Ensure that all exaggerated/subjective language is converted clearly to more neutral terms.

- Ensure that all of the text layout is converted properly to a format that is more scannable.
- Ensure that all of the text is as concise as possible.

10.4.4 Help for Users

Experience indicates that web users generally do not read web pages in a serial manner; rather, they hop from one visual element to another. Thus, the biggest challenge faced by web page designers is to use the visual elements most effectively to draw the attention of potential users to key elements. Furthermore, it should be emphasized that users will not read text unless they are specifically enticed to do so. The following list presents some important guidelines/pointers [14]:

- Ensure that the application of visual-highlighting approaches is consistent throughout the website.
- Ensure that the key text has the highest possible contrast.
- Past experience clearly indicates that items above and to the left of the page center appear to be noticed first.
- Reinforce the hierarchy of web page contents with a visual dominance hierarchy.
- Test the final design of each and every web page by eliminating all visual elements in question, one at a time.
- Use size to make users understand which elements fall where in regard to the content hierarchy.
- Past experience indicates that users generally assume that a row of similar elements should be "read" from left to right or from top to bottom.
- Use the same color to develop a common thread among elements that cannot be placed next to each other.

10.4.5 Font Usage

Fonts are used to create a variety of web page elements, including menus, navigation bars, footers and headers, links, buttons, and tables, in addition to the text that conveys most of the content of a website. Font faces fall under two basic classifications: sans-serif and serif. Sans-serif fonts are simpler in shape because they consist of only basic line strokes. Two examples of sans-serif fonts are Arial and Helvetica.

Serif fonts have small appendages at the bottoms and tops of letters. Three examples of such fonts are Times Roman, Century, and Courier. These fonts are useful as they make it easier to read long lines.

Some of the pointers directly or indirectly concerned with font usage are as follows [2, 14]:

- Use italics for defining terms or emphasizing an occasional word.
- Avoid specifying absolute font sizes and getting carried away in the use of font styles, faces, and sizes.
- Note that different browsers support different font faces.

10.5 Website Design

Generally, more attention is paid to page design than site design. However, from the usability point of view, the site design is more important and challenging. Some of the important usability dos and don'ts with respect to website design are presented in Table 10.2 [14].

Some of the important factors to be considered with care in website design are shown in Figure 10.2 [11, 14]. Each of the important factors shown in Figure 10.2 is described in the following subsections.

10.5.1 Shared Elements of Site Pages

In regard to the effective usability of a site, it is important to help potential users become familiar with the site with minimal effort on their part. This can be accomplished by adapting a consistent page style that repeats common elements throughout the website. This approach can also be helpful in improving user speed. Another approach that significantly improves usability is to concentrate all common elements at the bottom and top of each page or along the left-hand side.

TABLE 10.2

Website Design: Important Usability Dos and Don'ts

No.	Dos	Don'ts
1	Ensure that the web pages clearly honor the browser settings of the user.	Avoid having a banner page.
2	Ensure that all pages of a single website share a common look and feel.	Try not to pop up windows without the consent of the user.
3	Ensure that each and every web page incorporates real content.	Avoid saying "welcome" on web pages.
4	Ensure that web pages support browser resizing as much as possible.	Avoid having a copyright notice. It is not necessary for establishing ownership to web materials.
5	...	Avoid the use of frames.

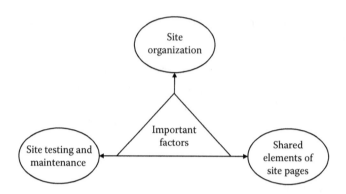

FIGURE 10.2
Some important factors to be considered in website design.

As per the work of Nielsen [11] and Brown [14], some of the user expectations of common elements are as follows:

- Incorporation of a help feature only for those situations in which it provides substantive information
- Capability to go to the home page by simply clicking on the icon of the website
- Incorporation of a search mechanism only when the site includes a rather large number of pages
- Gathering of information intended for all sponsoring agencies under "about us"
- Very clear display of a "contact us" mechanism

10.5.2 Site Testing and Maintenance

Regular testing and maintenance of the website are very important to maintain usability effectiveness. To maintain the quality of a website, test the design and each and every new page by using Internet Explorer and Netscape, disabling images, using different browser window widths, and making use of a dialup connection [14].

As the web pages tend to change over time, it is very important to conduct appropriate maintenance activities on a regular basis. At a minimum of once a month, all of the links that are still in active mode must be verified. Furthermore, whenever a web page is changed, every effort should be made to double check that its links are operating in a normal fashion.

10.5.3 Site Organization

Past experience indicates that site organization needs careful consideration during design because users generally do not read web pages the way they read books. The following list presents some useful guidelines concerning website organization [11, 14]:

- Organize the site into many bite-size pieces capable of being traversed in varying ways to take advantage of the navigational flexibility of the web.
- Do not display blocks of text in a large font.
- Provide users some content on each page.
- Ensure that the pertinent information is positioned in such a way that it is still visible even in a situation when the browser window is shrunk to around 50% of the screen width.
- Ensure that pointers to related topics are clearly visible somewhere in the upper half of the page in question.

10.6 Navigation Aids

There are various types of navigation aids employed by users to find their way around websites. Some of the important usability "dos" and "don'ts" concerning navigation aids are presented in Table 10.3 [14, 15].

Three important factors that need to be considered with care in regard to the navigation aids are shown in Figure 10.3 [2, 14, 15]. Each of the important factors shown in Figure 10.3 is described in the following subsections.

10.6.1 Menus and Menu Bar Usage

Normally, websites use menu bars to provide fundamental navigation-related functionality. They may incorporate various links or include menu titles that

TABLE 10.3

Navigation Aids: Important Usability Dos and Don'ts

No.	Dos	Don'ts
1	Ensure the proper conformance to the standard practice of having all links underlined	Do not change the standard colors used for links
2	Organize all navigation aids by considering the user tasks to be carried out	Do not label menu items, links, or buttons with meaningless phrases
3	Ensure that menu items, links, or navigational bar items that lead to the "current location" are deactivated in a proper manner	Do not confuse users by implementing menus in a "creative" way
4	Give proper importance to bread-crumb trails or navigation bars because they provide users with an understanding of location on a site	Do not assume that all users will be able to familiarize themselves with the website

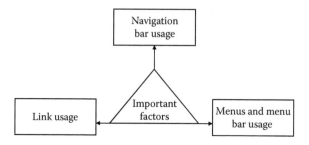

FIGURE 10.3
Three important factors to be considered in regard to navigation aids.

drop down as the user's cursor passes over them or clicks on them. The following list presents some useful guidelines for the use of menus and menu bars [2, 14, 15].

- Ensure that menu items are grouped together logically.
- Ensure that menu titles form a consistent group and are short.
- Do not use cascading (i.e., multilevel) menus.
- Ensure that all menus are anchored to a menu bar across the top of the web page.
- Format menu items and menu titles by utilizing lowercase and uppercase letters.

10.6.2 Navigation Bar Usage

The principal objective of having navigation bars is to lay out the website structure in a hierarchical form, and they are generally located along one side of the web page. The following two factors have proved to be quite helpful in using the navigation bar effectively [2]:

- The selection of the top ten things (during the navigation bar development process) that all users are likely to do on the site in question
- The selection of navigation labels and structure with utmost care

The following seven steps are considered very useful in selecting navigation labels and structure [14, 15]:

Step 1: Establish a list of functionalities/operations that the website has to support if potential users are to perform the identified top ten tasks/things effectively.

Step 2: Write down clearly the operations on individual index cards.

Step 3: Spread out the cards with care and group them into logical categories.

Step 4: Identify at least five users and have them repeat the preceding three steps.

Step 5: Compare the findings of all the sortings. If a pattern of categories/ classifications is not emerging, then repeat the whole process (i.e., all the steps) again.

Step 6: When the categories/classifications arrived at by most users look similar, take advantage of them to develop an outline of the website structure.

Step 7: Present the outline of the structure to at least five other users and ask for their inputs. Repeat the process as the need arises.

10.6.3 Link Usage

Links are probably the most common mechanism that supports website navigation, and web pages make use of links in the following three ways [2, 14, 15]:

- To direct users to pages that contain additional information on the graphic/text mentioned in the link
- To direct users to an alternative source when the current page does not contain the needed information
- To provide efficient access to the website's other pages

The following list presents some guidelines considered useful in regard to the effective use of links [14, 15]:

- Choose link text with utmost care.
- Make use of standard colors and underline all links.
- Do not underline the nonlink text.
- Make the image itself be the link when there is a definite need to link to a larger copy of an image.
- Format links by utilizing lowercase and uppercase letters.
- Group the links into classifications when multiple links show up in a list.
- Underline the words that really matter to improve the readability of the link.
- Locate the alternative links at the top of the page.

10.7 Tools for Evaluating Web Usability

There are many tools or methods that can be used to identify web usability problems. Over the years, these tools have proved to be quite useful in checking routine site-design elements in regard to consistency and in encouraging the application of good design practices. Four of these tools are as follows [2, 16]:

- NetRaker
- Web SAT
- Lift
- Max

Each of these four tools or methods is described in the following subsections.

10.7.1 NetRaker

NetRaker consists of a number of online tools that help to highlight usability problems and perform market research. NetRaker provides a set of comprehensive guidelines to compose objective survey questions and a customizable set of usability survey templates. The questions are randomly made available to the users of the website by providing them with an option to participate.

The survey requires users to perform tasks on the website and then provide satisfactory feedback in regard to the simplicity of carrying out the tasks. Some of the main benefits of NetRaker are as follows [2, 16]:

- This is a very useful method to survey users and gather usability-related feedback quickly.
- The NetRaker automation ensures that users are surveyed consistently.
- This is a useful tool to obtain feedback in the context of the intended purpose of a website as opposed to relying totally on generic hypertext markup language (HTML) checks or statistical analysis.

Finally, it is added that NetRaker is one of the best methods/tools to identify usability issues because it is based on users' direct feedback.

10.7.2 Web SAT

The Web Static Analyzer Tool (Web SAT) is used for checking web page html against typical usability guidelines for potential problems. It is one of the methods/tools that belongs to the Web Metrics Suite developed by the National Institute of Standards and Technology (NIST) [2]. Web SAT permits generally up to five individual Uniform Resource Locators (URLs) to be checked against its usability-related guidelines.

At the end of evaluation, Web SAT provides a comprehensive report of problems discovered on each web page entered. The problems are grouped under the following six classifications [2, 16]:

Classification I: Performance. Under this classification, the problems are concerned with the size and coding of graphics with respect to page download speeds.

Classification II: Form use. Under this classification, the problems are concerned with the form Submit and Reset buttons.

Classification III: Navigation. Under this classification, the problems are concerned with the coding of links.

Classification IV: Accessibility. Under this classification, the problems are concerned with the page making proper use of tags for visually impaired users.

Classification V: Readability. Under this classification, the problems are concerned with content readability.

Classification VI: Maintainability. Under this classification, the problems are associated with tags and coding information that would make the page easier to port to another server.

All in all, the principal limitation of the Web SAT is that it can examine or check only individual web pages.

10.7.3 Lift

Lift is another usability tool used to perform analysis of a web page to uncover usability-related problems. There are two types of Lift [2].

Lift Onsite. This can be easily run from a personal computer (PC), and it provides the very compelling feature of directly fixing the HTML-related problems as they are being reviewed in the usability evaluation report.

Lift Online. This conducts HTML checks derived from usability principles in a similar way to Web SAT. More specifically, it checks one page at a time and then provides a report on the usability issues of a page. Furthermore, Lift Online goes a step further than Web SAT because it provides appropriate code change recommendations.

All in all, Lift provides usability-based HTML validations to ensure good coding practices.

10.7.4 Max

This is another useful usability tool that scans through a website to collect information about vital statistics and rates concerning a site's usability.

Max uses a statistical model to simulate the experience of a user in calculating ratings in the following three areas [2, 16]:

Content. Max summarizes the percentage of different media elements (i.e., text, graphics, and multimedia) in addition to client-side technologies used (e.g., Flash and Portable Document Format [PDF]) that comprise the website.

Load time. In this case, Max estimates the mean time to load website pages.

Accessibility. Max estimates the mean time a user takes to find something on the site under consideration.

The main weakness of Max is that it does not provide many suggestions to make changes to design. Its main strength is that it provides a performance benchmark.

10.8 Questions to Evaluate Effectiveness of Website Message Communication

This section presents a checklist of questions considered useful in evaluating the effectiveness of messages communicated by a website. These questions will be helpful (directly or indirectly) in improving web usability, and they are grouped under the following six distinct areas [4, 17–19].

- Text
- Concept
- Content
- Mechanics
- Navigation
- Design

Questions pertaining to each of these six areas are presented in the following subsections.

10.8.1 Text

Some of the questions belonging to this area are as follows [4, 17–19]:

- Is the text on the first and feature pages short enough for its effective application?
- Is the text sufficiently attractive for reading?

- Are the hyperlink and button titles clear and straightforward?
- Are the titles and subheadings informative enough for their effective application?
- Is the text under consideration grammatically checked?
- Are all of the titles appropriate for effective use by search engines?
- Does the first page convey a clear message to potential visitors, including what they can expect to find in the website?
- Can the text be read properly in a cursory manner?

10.8.2 Concept

Some of the questions belonging to this area are as follows [2, 17–19]:

- What expectations will the website raise for the visitors?
- Can the rest of the site satisfy this promise properly?
- What basic image of the organization/company does the site project?
- What existing websites can be compared with the one in question or under consideration?
- What does the first page clearly promise concerning the rest of the website?

10.8.3 Content

Some of the questions belonging to this area are as follows [2, 17–19]:

- Is the content easily accessible and the purchasing procedure (if applicable) clearly user friendly?
- How accurate and unique are the website contents?
- What information is the website expected to convey to its users?
- What portion of the web page(s) was allocated for content in comparison to other factors?
- Does the site provide any communication-related options?
- Is the information on the website accessible, clear, accurate, attractive, and complete?

10.8.4 Mechanics

Some of the questions belonging to this area are as follows [2, 17–19]:

- How quickly does the site react; how quickly do the pages load?
- Do tools such as roll-down menus and mouseover events (if utilized) clearly support the use of the site?

- Are there any error-related messages?
- Are buttons and hyperlinks operating as per requirements?
- How functional is the website under consideration?

10.8.5 Navigation

Some of the questions belonging to this area are as follows [2, 17–19]:

- Will the browsing location be clear to users?
- Is it possible to properly predict the contents of the options on the menu without clicking them?
- Is it possible to find anything using simple keywords on the engine?
- How does the site conform to existing web standards?
- Does the site contain a local search engine?
- How simple and straightforward is it to use the hyperlinks, buttons, and menu to browse the site?

10.8.6 Design

Some of the questions belonging to this area are as follows [2, 17–19]:

- Is the style of the site unique?
- Is the design of the site clearly impressionable?
- Is there an appropriate level of contrast between text color and background?
- Can the design in any way deter potential users?
- Are the users directed properly to the most important page elements?
- Is there proper balance between content and design?

Problems

1. List at least seven facts and figures concerned with web usability.
2. Define "Web usability" and write an essay on web usability.
3. List and discuss commonly occurring web design errors.
4. List at least five web page design usability dos and don'ts.
5. What are the important factors to be considered in web page design?
6. List at least four website design-related usability dos and don'ts.
7. Discuss at least two factors to be considered in website design.

8. List at least four navigation-aids-related usability dos and don'ts.
9. Describe the following two tools used to evaluate web usability:

   a. Web SAT

   b. NetRaker

10. Write an essay on questions to evaluate the effectiveness of the website message: Communication.

References

1. Powell, T. (2000). *Web design: The complete reference*. Berkeley, CA: Osborne McGraw-Hill.
2. Dhillon, B. S. (2004). *Engineering usability: Fundamentals, applications, human factors, and human error*. Stevenson Ranch, CA: American Scientific Publishers.
3. Cloyd, M. H. (2001). Designing user-centered Web applications in Web time. *IEEE Software, 18*(1), 62–69.
4. McDonald, S., Waern, Y., & Cockton, G. (Eds.). (2000). *People and computers XIV: Usability or else!* London: Springer-Verlag.
5. Preece, J. (2000). *Online communities: Design usability, supporting sociability*. New York, NY: John Wiley and Sons.
6. Becker, S. A., & Mottay, F. E. (2001). A global perspective on Web site usability. *IEEE Software, 18*(1), 54–61.
7. Chi, E. H. (2002). Improving Web usability through visualization. *IEEE Internet Computing, 6*(2), 64–71.
8. Trenner, L., & Bawa, J. (Eds.). (1998). *The politics of usability: A practical guide to designing usable systems in industry*. London: Springer-Verlag.
9. Souza, R. K., Manning, H., Goldman, H., & Tong, J. (2000, October). *The best of retail site design* (White paper). Cambridge, MA: Forrester Research.
10. Nielsen, J. (1996). *Top ten mistakes in Web design*. Retrieved December 10, 2012, from Alertbox Web site: www.useit.com/alertbox/9605.html
11. Nielsen, J. (2000). *Designing Web usability*. Indianapolis, IN: New Riders Publishing.
12. Manning, H., McCarthy, J. C., & Souza, R. K. (1998, September). *Why most Web sites fail* (White paper). Cambridge, MA: Forrester Research.
13. Nielsen, J. (2003). *PR on websites: Increasing usability*. Retrieved December 10, 2012, from Alertbox Web site: www.useit.com/alertbox/20030310.html
14. Brown, G. E. (2003). *Web usability guide*. Richmond, CA: NEES Consortium. Retrieved December 10, 2012, from NEEShub Web site: www.nees.org/info/contact-us.html
15. Fleming, J. (1998). *Web navigation: Designing the user experience*. Sebastopol, CA: O'Reilly and Associates.
16. Chak, A. (2000, August). *Usability tools: A useful start* (New Architect: Strategy product review). Retrieved December 10, 2012, from www.webtechniques.com/archives/2000/08/stratevu/

17. Price, J., & Price, L. (2002). *Hot text: Web writing that works.* Indianapolis, IN: New Riders Publishing.
18. Niederst, J. (2001). *Web design in a nutshell: A desktop quick reference.* Sebastopol, CA: O'Reilly and Associates.
19. Williams, R., & Tollett, J. (2000). *The non-designer's Web book: An easy guide to creating, designing and posting your own website.* Berkeley, CA: Peachpit Press.

11

Computer System Life-Cycle Costing

11.1 Introduction

Computer systems are made up of both hardware and software parts, and over the years, the percentage of overall computer system cost spent on the hardware part has changed quite dramatically. For example, in 1955 the hardware part accounted for about 80% of overall computer system cost; by 1980, the cost of the hardware part had decreased to around 10% [1, 2]. Nowadays, the cost of software is a very important component of the overall computer system cost. More specifically, the cost of software plays a crucial role in the computer system life-cycle cost.

Over the years, significant effort has been devoted to develop models and procedures to estimate computer system hardware cost, software cost, and life-cycle cost. As the result of this effort, many models and procedures have been developed to estimate, directly or indirectly, such costs. This chapter presents various important aspects of computer system life-cycle costing.

11.2 Models for Estimating Computer System Life-Cycle Cost

There are a number of mathematical models that can be used to estimate the life-cycle cost of a computer system. Two of these models are presented here [2–4].

11.2.1 Model I

This model divides the life-cycle cost of a computer system into two parts: acquisition cost and ownership cost. Thus, the computer system life-cycle cost for Model 1 ($CSLC_{C1}$) is defined as

$$CSLC_{C1} = A_{C1} + O_{C1} \tag{11.1}$$

where

$\quad$ $CSLC_{C1}$ = computer system life-cycle cost for model 1

$\quad$ A_{C1} = acquisition cost of the computer system for model 1

$\quad$ O_{C1} = ownership cost of the computer system for model 1

The acquisition cost includes the cost of items such as documentation, installation, software license fees, training, and system hardware. Similarly, the ownership cost includes the cost of items such as computer system downtime, supplies, corrective maintenance, and preventive maintenance.

Example 11.1

An organization using a computer system for certain purposes is considering replacing it with a better one. Two different computer systems are being considered for its replacement, and their data are presented in Table 11.1. Determine which of the two computer systems should be purchased to replace the existing computer system in regard to their life-cycle costs.

$\quad$ The annual expected failure costs (AF_C) of computer systems A and B using Table 11.1 data are given by

$$AF_{C-A} = (0.01)(3,000) = \$30$$

and

$$AF_{C-B} = (0.02)(2,000) = \$40$$

where

$\quad$ AF_{C-A} and AF_{C-B} = the annual expected failure costs of computer systems A and B, respectively

Using the given and calculated data in an equation taken from Dhillon [2], we obtain the following present values of failure cost (PF_C) over the useful lives of computer systems A and B:

$$PF_{C-A} = 30 \left[\frac{1-(1+0.04)^{-15}}{0.04} \right]$$

$$= \$333.55$$

TABLE 11.1

Data for Two Computer Systems Under Consideration

No.	Description	Computer System A	Computer System B
1	Acquisition cost	$200,000	$150,000
2	Expected useful life in years	15	15
3	Cost of a failure	$3,000	$2,000
4	Annual failure rate	0.01	0.02
5	Annual operating cost	$2,500	$7,000
6	Annual interest rate	4%	4%

and

$$PF_{C-B} = 40 \left[\frac{1-(1+0.04)^{-15}}{0.04} \right]$$

$$= \$444.73$$

where
  PF_{C-A} and PF_{C-B} = the present values of failure costs over the useful lives of computer systems A and B, respectively.

Similarly, using the given data in the equation taken from Dhillon [2], we obtain the following present values of operating costs (PO_C) over the useful lives of computer systems A and B:

$$PO_{C-A} = [2,500] \left[\frac{1-(1+0.04)^{-15}}{0.04} \right]$$

$$= \$27,795.96$$

$$PO_{C-B} = [7,000] \left[\frac{1-(1+0.04)^{-15}}{0.04} \right]$$

$$= \$77,828.71$$

where
  PO_{C-A} and PO_{C-B} = the present values of operating costs over the useful lives of computer systems A and B, respectively

The life-cycle costs (LC_C) of computer systems A and B are

$$LC_{C-A} = 200,000 + 333.55 + 27,795.96$$

$$= \$228,129.51$$

and

$$LC_{C-B} = 150,000 + 444.73 + 77,828.71$$

$$= \$228,273.44$$

where
  LC_{C-A} and LC_{C-B} = the life-cycle costs of computer systems A and B, respectively

The life-cycle cost of computer system A is lower, and thus it should be purchased.

11.2.2 Model II

This is a more detailed model to estimate computer system life-cycle cost, but it assumes that the corrective maintenance cost is the only ownership cost of the computer system. Thus, the computer system life-cycle cost for Model 2 ($CSLC_{C2}$) is defined as

$$CSLC_{C2} = A_{C2} + \sum_{j=1}^{m} CM_{Cj}\theta_j/(1+i)^j \qquad (11.2)$$

where
  $CSLC_{C2}$ = computer system life-cycle cost for Model 2
  A_{C2} = acquisition cost of the computer system for Model 2
  m = expected life of the computer system, expressed in years
  i = discount rate
  θ_j = expected number of times the computer system fails in year j
  CM_{Cj} = single corrective maintenance call cost during year j

When the expected number of computer system failures occurring for each year is the same, Equation (11.2) becomes

$$CSLC_{C2} = A_{C2} + \theta \sum_{j=1}^{m} CM_{Cj}/(1+i)^j \qquad (11.3)$$

where
  θ = expected number of computer system failures per year

Example 11.2

Assume that the acquisition cost of a computer system is $7,000 and that its expected useful life is 7 years. The expected number of the computer system failures is 500 failures/million hours. The only ownership cost involved is the cost of corrective maintenance and is $600 per corrective maintenance call. Estimate the life-cycle cost of the computer system if the annual discount or interest rate is 5%.

The expected number of computer system failures per year is given by

$$\theta = \frac{(8,760)(500)}{1,000,000} = 4.38 \text{ failures/year}$$

By substituting this calculated value and the given data values into Equation (11.3), we get

$$CSLC_{C2} = 7,000 + (4.38)(600)\sum_{j=1}^{7} \frac{1}{(1+0.05)^j}$$

$$= \$22,206.39$$

Thus the estimated life-cycle cost of the computer system is $22,206.39.

11.3 Models for Estimating Computer System Servicing-Labor Cost and Maintenance Cost

Over the years, a number of mathematical models have been developed to estimate the costs concerned with computer system maintenance [2–4]. This section presents two such models: One model is concerned with estimating the cost of labor to service a computer system, and the other model addresses maintenance cost for the computer system.

11.3.1 Model I: Computer System Servicing-Labor Cost

This model is concerned with estimating the yearly cost of labor to service a computer system. This cost depends on factors such as follows [2–4]:

- Mean time to repair
- Mean time between failures
- Average cost of labor
- Mean time to preventive maintenance
- Preventive maintenance time interval

The annual labor cost (AL_C) of servicing a computer system is expressed by

$$AL_C = (HL_C)(8,760)\left[\frac{(PM_{MTTP} + PM_{TT})}{PM_{MTBS}} + \frac{(MTTR + CM_{TT})}{MTBF}\right] \quad (11.4)$$

where
  AL_C = annual labor cost of servicing a computer system
  HL_C = hourly labor cost
  $8{,}760$ = number of hours in a year
  PM_{MTTP} = mean time to perform preventive maintenance
  PM_{TT} = average travel time associated with a preventive maintenance
  PM_{MTBS} = mean time between preventive maintenance services
  $MTTR$ = mean time to repair
  CM_{TT} = average travel time associated with a corrective maintenance call
  $MTBF$ = mean time between failures

> **Example 11.3**
>
> Assume that we have the following data concerning servicing a computer system in an organization:
> $MTTR = 4$ hours
> $MTBF = 5{,}000$ hours
> $HL_C = \$40$
> $PM_{TT} = 0.5$ hour

$\text{PM}_{\text{MTTP}} = 3 \text{ hours}$
$\text{PM}_{\text{MTBS}} = 1{,}500 \text{ hours}$
$\text{CM}_{\text{TT}} = 2 \text{ hours}$

Calculate the annual labor cost for servicing the computer system with the aid of Equation (11.4).

By inserting the specified data values into Equation (11.4), we obtain

$$\text{AL}_C = (40)(8{,}760)\left[\frac{(3+0.5)}{1{,}500} + \frac{(4+2)}{5{,}000}\right]$$

$$= \$1{,}238.08$$

Thus, the annual labor cost for servicing the computer system is $1,238.08.

11.3.2 Model II: Computer System Maintenance Cost

This model is concerned with estimating the monthly cost of computer system hardware maintenance, which is defined by [2–5]

$$\text{HM}_C = \text{I}_C + \text{CM}_C + \text{PM}_C \tag{11.5}$$

where
  HM_C = hardware maintenance cost
  I_C = inventory cost
  CM_C = corrective maintenance cost
  PM_C = preventive maintenance cost

The inventory cost, I_C, is defined by

$$\text{I}_C = \alpha(\text{V}_M) \tag{11.6}$$

where
  α = monthly inventory cost rate, which includes interest charges for spares, depreciation, handling cost, etc.
  V_M = value of maintenance spare parts inventory

The corrective maintenance cost (CM_C) for the computer system hardware is expressed by

$$\text{CM}_C = \beta(\text{EOH})(\text{CM}_{\text{TT(CE)}} + \text{MTTR})/\text{MTBF} \tag{11.7}$$

where
  β = customer engineer's hourly rate, including the usage rate of spare parts
  EOH = equipment operating hours per month

$CM_{TT(CE)}$ = travel time of customer engineer to perform corrective maintenance
MTTR = mean time to repair
MTBF = mean time between failures

The customer engineer's hourly rate, β, is defined by

$$\beta = \frac{HR_{CE}(1+OR)}{\mu} + P_C \qquad (11.8)$$

where
HR_{CE} = hourly rate for customer engineer
OR = overhead rate
μ = fraction of time that the customer engineer spends on the maintenance activity (Note that the remaining fraction includes time he/she spends on items such as waiting, paperwork, and training.)
P_C = cost of parts per hour

The computer system hardware preventive maintenance cost, PM_C, is defined by

$$PM_C = \beta(EOH)(PM_{ST(CE)} + PM_{TT(CE)})/PM_{SI} \qquad (11.9)$$

where
$PM_{ST(CE)}$ = preventive maintenance, scheduled time of the customer engineer
$PM_{TT(CE)}$ = preventive maintenance, travel time of the customer engineer
PM_{SI} = preventive maintenance, scheduled interval

Additional information on this model is available in the literature [2–5].

11.4 Software Costing and Associated Problems and Steps

Over the years, the software cost component of the total computer system cost has increased to a very high level from its beginning as a rather low percentage of the total computer system cost. For example, as per a U.S. Air Force study performed in 1972, the software cost, in 1955, accounted for less than 20% of the combined total of the computer system hardware and software costs, and its projection for 1985 was about 80% of the total amount [2, 4]. In addition, in July 1976, *Newsweek* magazine stated that the ratio of computer system hardware cost to software cost was 1 to 4 in 1976 compared to 4 to 1 in the 1950s.

This drastic change has emphasized the need for careful attention to estimating software cost, as this has become a very important element in the total computer system life-cycle cost. Over time, many models and methods have been developed to estimate software costs. Needless to say, one faces many problems in estimating software costs because it is a complex process. Some of these problems are as follows [2, 4, 6]:

- Poor comprehension of the software development process and maintenance
- Shortage of adequate historic data for making appropriate checks
- Poor understanding of the management- and technical-related constraints effects
- Inhibition of project-to-project comparisons because of firm belief in a project's uniqueness
- Unavailability of sufficient historic data for calibration purposes (A calibration may simply be described as a process through which a model is fitted to a specified cost estimating situation.)

There are a number of steps involved in a software costing process. The basic ones are shown in Figure 11.1 [7]. The steps shown in the figure [4, 7] are

1. Define objectives
2. Make plan for necessary required resources and data
3. Highlight precise software requirements
4. Carry out detail as possible
5. Use several independent cost estimation methods and sources
6. Compare and iterate end results
7. Take follow-up appropriate actions

As many of these steps are considered self-explanatory, only some of them are discussed or described here.

In regard to Step 1 (define objectives), it will be very useful to pay close attention to the following guidelines [4, 7]:

- Key the objective to the requirements for decision-making information.
- Balance the objective of estimating accuracy for the various system elements of the cost estimates.
- Review objectives as the process progresses.
- Make changes as needed.

In regard to Step 4 (carry out detail as possible), one is basically concerned with working out as much detail as possible. Broadly speaking, one may simply say that the more detail or depth to which one carries out cost estimation, the more accurate will be the final results.

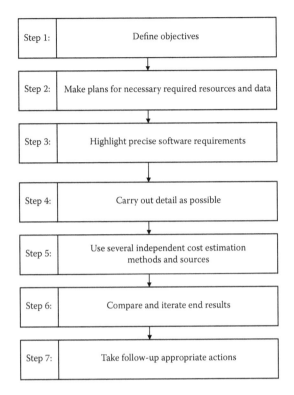

Step 1:	Define objectives
Step 2:	Make plans for necessary required resources and data
Step 3:	Highlight precise software requirements
Step 4:	Carry out detail as possible
Step 5:	Use several independent cost estimation methods and sources
Step 6:	Compare and iterate end results
Step 7:	Take follow-up appropriate actions

FIGURE 11.1
Basic steps of a software costing process.

In regard to Step 5 (use several independent cost estimation methods and sources), the main reason for making use of a combination of cost estimating methods or approaches is to avoid the weak points of a single method and to take advantage of their combined strength. Some examples of the cost estimating methods are expert opinions as well as top-down, bottom-up, algorithmic, and analogy models.

Additional information on the remaining steps shown in Figure 11.1 is available in a work by Boehm [7].

11.5 Model for Estimating Software Life-Cycle Cost and Influencing Factors

The life-cycle cost of software is made up of seven main elements and is defined by [2, 8]

$$SLC_C = C_1 + C_2 + C_3 + C_4 + C_5 + C_6 + C_7 \tag{11.10}$$

where

SLC$_C$ = software life-cycle cost

C_1 = software design cost made up of data structure cost, cost of flow charts, cost of test procedures, and cost of input and output parameters

C_2 = software analysis cost made up of cost of system requirements, cost of program requirements, cost of interface requirements, and cost of design requirements and specifications

C_3 = software code and checkout cost made up of cost of desk checks, cost of coded instructions, and cost of compiling programs

C_4 = software operating and support cost made up of cost of modifications, cost of documentation revisions, cost of test revisions, and cost of environments

C_5 = software installation cost made up of verification cost, validation cost, and certification cost

C_6 = software test and integration cost made up of cost of program test and cost of system integration

C_7 = software documentation cost made up of cost of user manual, cost of listings, and cost of maintenance manual

There are many factors that influence software life-cycle cost. They may be grouped under five classifications, as shown in Figure 11.2 [9]. Classification I contains product attributes whose examples are required software reliability, software product complexity, and the choice of programming language. Classification II contains project attributes whose

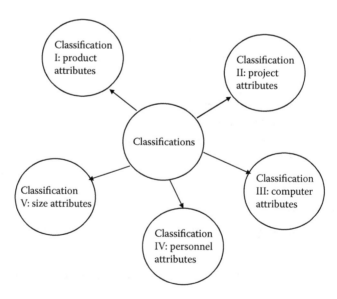

FIGURE 11.2
Classifications of factors influencing software life-cycle cost.

examples are schedule constraints, modern programming practices, and use of software tools. Classification III contains computer attributes whose examples are turnaround time, speed, and storage constraints. Classification IV contains personnel attributes that affect software cost much more than any other classifications of attributes. Some examples of the personnel attributes are teamwork, personnel and team capabilities, and experience in regard to items such as programming language, applications, and virtual machines. Finally, Classification V contains size attributes whose examples are number of inputs, outputs, instructions, and data elements.

11.6 Models for Estimating Software Costs

Over the years, various models have been developed to estimate software costs [4, 6, 7, 10]. Some of these models are presented here.

11.6.1 Model I

This model is concerned with estimating the software quality cost, SQ_C, which is defined by [11]

$$SQ_C = C_a + C_p + C_i + C_e \tag{11.11}$$

where
SQ_C = software quality cost
C_a = appraisal cost associated with tasks or activities concerned with detecting software errors in software projects under consideration (The components of the appraisal cost include the cost of reviews, the cost of software testing, and the cost of ensuring the quality of external participants such as subcontractors.)
C_p = prevention cost associated with tasks or activities concerned with preventing the occurrence of software errors (Some examples of these tasks or activities are developing a software quality infrastructure, making improvements and updating the infrastructure, and carrying out the regular tasks or activities essential for its successful functioning.)
C_i = internal failure cost associated with the eradication of software errors found during design reviews, testing, and acceptance tests prior to the installation of the software at customer sites
C_e = external failure cost associated with the eradication of software errors found by customers after the installation of software at their sites

11.6.2 Model II

This model is concerned with estimating the annual software marketing cost, ASM_C, which is defined by [12]

$$ASM_C = HO_C + FS_C \tag{11.12}$$

where
   ASM_C is the annual software marketing cost
   HO_C = home office cost
   FS_C = field sales-related cost

The field sales-related cost, FS_C, is expressed by

$$FS_C = [\alpha(SS) + \alpha(SA)n](1+OR) + CR(APS) \tag{11.13}$$

where
   α = number of people involved in sales
   SS = annual base salary of a salesperson
   SA = system analyst's annual salary
   n = number of system analysts employed per salesperson
   OR = overhead rate
   CR = commission rate
   APS = annual product sales

11.6.3 Model III

This model is concerned with estimating the software development cost, which is defined by [13, 14]

$$SD_C = SPD_C + SSD_C \tag{11.14}$$

where
   SD_C = software development cost
   SPD_C = software primary development cost
   SSD_C = software secondary development cost

The software secondary development cost, SSD_C, is defined by

$$SSD_C = \theta SPD_C$$

$$= \sum_{j=1}^{n} C_j \tag{11.15}$$

where
θ = ratio of software secondary development cost to software primary development cost
n = number of secondary resources
C_j = cost of secondary resource j, for j = 1,2,3,...,n

Similarly, the software primary development cost, SPD_C, is expressed by

$$SPD_C = (mm)R \qquad (11.16)$$

where
mm = manpower needed for software development, expressed in man-months (It includes activities such as debug, code, analysis, design, test, and checkout.)
R = software development manpower average labor rate expressed in dollars per man-month (It includes costs such as administration cost, overhead cost, and general cost.)

Additional information on this model is available in the literature [13, 14].

11.6.4 Model IV

This model is concerned with estimating software project-related effort, in programmer-months, when little information concerning the project under study is available, except its expected delivery instructions. The software project-related effort, E_{sp} is defined by [15]

$$E_{sp} = (1.7)\beta\mu k \qquad (11.17)$$

where
E_{sp} = software project-related effort expressed in programmer-months
β = software complexity factor (The values of this factor for very complex, moderately complex, and trivial software are 10, 5, and 1, respectively.)
μ = labor estimate adjustment factor expressed in decimal fraction (The recommended values of this factor for poorly managed projects and under best conditions are 2.9 and 0.435, respectively.)
k = delivered instructions expressed in thousands

Additional information on this model is available in the work by Schneider [15].

11.6.5 Model V

This model is concerned with estimating a software project's minimum duration, under the assumption that the entire hardware will be available during the project life. Thus, the minimum software project's duration is defined by [15]

$$T_m = T_p/S_a \qquad (11.18)$$

where
  T_m = minimum software project's duration
  T_p = total programmer-months
  S_a = average staff size allocated to the software project

Additional information on this model is available in the work by Schneider [15].

11.7 Models for Estimating Software Costs

There are many methods used to estimate software costs, including the ones shown in Figure 11.3 [4, 7]. These methods are algorithmic models, top-down estimating, expert opinion, analogy, and bottom-up estimating. The algorithmic models are described here, and information on the other four methods is available in the literature [4, 7].

The algorithmic models may simply be described as the ones that provide at least one mathematical algorithm for generating a computer software

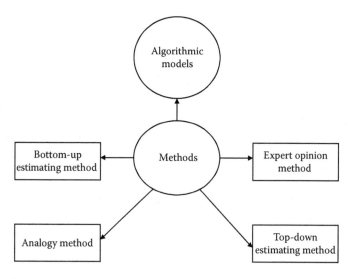

FIGURE 11.3
Software cost estimation methods.

cost estimate as a function of several variables. These variables are the important cost drivers. The five common types of algorithmic models are described in the following subsections [4, 7].

11.7.1 Multiplicative Models

These cost estimating models take the following form:

$$\text{EFT} = B_0 \prod_{i=1}^{k} B_i^{z_1} \tag{11.19}$$

where
  EFT = effort
  k = number of cost driver variables
  z_i = cost driver variable i, for $i = 1,2,3,\ldots,k$
  B_i = coefficient selected to best fit the observational data, for $i = 0,1,2,3,\ldots,k$

Past experience clearly indicates that this type of model works fairly well for reasonably independent chosen variables. Furthermore, to obtain development cost, one usually multiplies constant labor cost with the effort (EFT) value. Additional information on multiplicative models is available in the literature [14, 16].

11.7.2 Analytic Models

These cost-estimating models take the following form:

$$\text{EFT} = f(y_1, y_2, y_3, \ldots, y_k) \tag{11.20}$$

where
  EFT = effort
  k = number of cost driver variables
  f = function (this function is neither linear or multiplicative)
  y_j = cost driver variable j, for $j = 1,2,3,\ldots,k$

Two examples of the analytic models are the Halstead [17] and Putnam [18] models.

11.7.3 Linear Models

These cost-estimating models take the following form:

$$\text{EFT} = A_0 + \sum_{j=1}^{m} A_j y_j \tag{11.21}$$

where
EFT = effort
m = number of cost driver variables
y_j = cost driver variable j, for j = 1,2,3,...,m
A_j = coefficient selected to best fit the observational data, for j = 0,1,2,3,...,m

The history of the first usage of linear models goes back to the System Development Corporation Software cost-estimation study conducted in the mid-1960s [16]. Nonetheless, it is to be noted that there are many nonlinear interactions in software development for linear models if they are to perform effectively.

11.7.4 Tabular Models

These models are composed of tables relating values of the cost driver variables either to multipliers employed to adjust the effort estimate or to portions of the software development effort. Generally, tabular models are quite straightforward to understand and implement. Two examples of these models are the Boeing model [19] and the Wolverton model [20].

11.7.5 Composite Models

These cost estimating models incorporate an amalgamation of four types of functions (i.e., multiplicative, analytic, linear, and tabular) to determine software effort as a function of cost driver variables. The models (i.e., composite) are relatively more difficult to use and learn and require more effort and data. Two examples of composite models are the RCA PRICE-S [21] and TRW SCEP [22] models.

Problems

1. Discuss problems related to software costing.
2. Write an essay on computer system life-cycle costing.
3. Assume that the procurement cost of a computer system is $9,000 and that its expected useful life is ten years. The expected number of the computer system failures is 600 failures/million hours. The only ownership cost involved is the cost of corrective maintenance, which is $400 per corrective maintenance call.

   Calculate the life-cycle cost of the computer system, if the annual discount or interest rate is 4%.
4. What are the basic steps involved in a software costing process?
5. List at least four software cost estimation methods.

6. Discuss the classifications of factors influencing the life-cycle cost of software.

7. What are the algorithmic models? Describe at least two such models.

8. Define a model that can be used to estimate software quality cost.

9. Define a model that can be used to estimate software development cost.

10. Assume that we have the following data concerning the servicing of a computer system in an organization:

Computer system mean time to repair = 5 hours

Computer system mean time between failures = 6,000 hours

Labor cost per hour = $50

Average travel time associated with a preventive maintenance call = 0.6 hour

Mean time to perform preventive maintenance = 4 hours

Mean time between preventive maintenance services = 1,600 hours

Average travel time associated with a corrective maintenance call = 1.5 hours.

Calculate the annual labor cost for servicing the computer system by using Equation (11.4).

References

1. Keene, S. J. (1992). Software reliability concepts, *Annual Reliability and Maintainability Symposium Tutorial Notes*, 1–21.
2. Dhillon, B. S. (2010). *Life cycle costing for engineers*. Boca Raton, FL: CRC Press.
3. Phister, M. (1978). Analyzing computer technology costs: Part II, maintenance. *Computer Design*, *1978*(October), 109–118.
4. Dhillon, B. S. (1989). *Life cycle costing: Techniques, models, and applications*. New York: Gordon and Brach Science Publishers.
5. Dhillon, B. S. (1987). *Reliability in computer system design*. Norwood, NJ: Ablex Publishing.
6. Stanley, M. (1982). Software cost estimating (Royal Signals and Radar Establishment Memorandum No. 3472). Procurement Executive, Ministry of Defense, Malvern, Worcs., UK.
7. Boehm, B. W. (1981). *Software engineering economics*. Englewood Cliffs, NJ: Prentice-Hall.
8. Earles, M. E. (1981). *Factors, formulas, and structures for life cycle costing*. Concord, MA: Eddin-Earles.
9. Boehm, B. W. (1984). Software life cycle factors. In C. R. Vick & C. V. Ramamoorthy (Eds.), *Handbook of software engineering*. New York, NY: Van Nostrand Reinhold.

10. James, T. G. (1977). Software cost estimating methodology. *Proceedings of the IEEE National Aerospace and Electronics Conference, 22–28.*
11. Galin, D. (2004). *Software quality assurance.* Harlow, Essex, England: Pearson Education.
12. Phister, M. (1976). *Data processing technology and economics.* Santa Monica, CA: Santa Monica Publishing.
13. Doty, D. L., Nelson, P. J., & Stewart, K. R. (1977). *Software cost estimation study* (Report No. RADC-TR-77-220). Vol. II. Prepared by Doty Associates, Inc., Rockville, MD.
14. Herd, J. R., Postak, J. N., Russell, W. E., & Stewart, K. R. (1977). *Software cost estimation study* (Report No. RADC-TR-77-220). Vol. I. Prepared by Doty Associates, Inc., Rockville, MD.
15. Schneider, V. (1978). Prediction of software effort and project duration: Four new formulas. *Sigplan Notices, 1978*(13), 49–59.
16. Wlaston, C. E., & Felix, C. P. (1977). A method of programming measurement and estimation. *IBM Systems Journal, 16,* 54–73.
17. Halstead, M. H. (1977). *Elements of software science.* New York, NY: Elsevier.
18. Putnam, L. H. (1978). A general empirical solution to the macro software sizing and estimating problem. *IEEE Transactions on Software Engineering, 4,* 345–61.
19. Black, R. K. D., Curnow, R. P., Katz, R., & Gray, M. D. (1977). *BCS software production data* (Report No. RADC-TR-77-116). Boeing Computer Services (BCS). Available from the National Technical Information Services (NTIS), Springfield, VA.
20. Wolverton, R. W. (1974). The cost of developing large-scale software. *IEEE Transactions on Computers, 23,* 615–36.
21. Freiman, F. R., & Park, R. E. (1979, October). PRICE software model—Version 3: An overview. *Proceedings of the IEEE Workshop on Quantitative Software Models,* 32–41.
22. Boehm, B. W., & Wolverton, R. W. (1980). Software cost modeling: Some lessons learned. *Journal of Systems and Software, 1,* 195–201.

Appendix: Bibliography of Literature on Computer System Reliability, Safety, and Usability

A.1 Introduction

Over the years, a large number of publications on computer system reliability, safety, and usability have appeared in the form of journal articles, conference proceedings articles, books, technical reports, and so on. This appendix presents an extensive list of publications related to computer system reliability, safety, and usability as a resource for readers wishing to obtain additional information. The period covered by the listing is from 1967 to 2011.

A.2 Publications

Abernethy, C. N. (1988). Human-computer interface standards: Origins, organizations and comment, *International Review of Ergonomics*, 2, 31–54.

Adolf, J. A., & Holden, K. L. (1996). Touchscreen usability in microgravity. *Proceedings of the Conference on Human Factors in Computing Systems*, 67–68.

AFR-122-9. (1987, August). Nuclear surety design certification for nuclear weapon system software and firmware. Washington, DC: Department of the Air Force.

Al-Janabi, A., & Aspinwall, E. (1996). Using quality design metrics to monitor software development. *Quality World, 1996*(March), 25–34.

Amir, Y., Caudy, R., Munjal, A., Schlossnagle, T., & Tutu, C. (2003). N-way fail-over infrastructure for reliable servers and routers. *Proceedings of the IEEE International Conference on Dependable Systems and Networks*, 130–135.

Ammann, P. E., & Knight, J. C. (1988). Data diversity: An approach to software fault tolerance. *IEEE Transactions on Computers, 37*(4), 418–425.

Ammar, H. H., Cukic, B., Mili, A., & Fuhrman, C. (2000). A comparative analysis of hardware and software fault tolerance: Impact on software reliability engineering. *Annuals of Software Engineering, 10*, 103–150.

Anderson, J., Fleek, F., Garrity, K., & Drake, F. (2001). Integrating usability techniques into software development. *IEEE Software, 18*(1), 46–53.

Anderson, J. E., & Macri, F. J. (1967). Multiple redundancy applications in a computer. *Proceedings of the Annual Symposium on Reliability,* 553–562.

Anderson, J. R., & Jeffries, R. (1985). Novice LISP errors: Undetected losses from working memory. *Human Computer Interaction, 1,* 107–131.

Anderson, J. W., & Browne, J. C. (1976). Graph models of computer systems application to performance evaluation of an operating system. *Proceedings of the International Symposium on Computer Performance, Modelling, Measurement, and Evaluation,* 166–178.

Anjard, R. (1995). Software quality assurance considerations. *Microelectronics and Reliability, 35*(6), 995–1000.

Anonymous. (1971, September 20). Blow balloons. *Aviation Week Space Technology,* p. 17.

Anonymous. (1992). Total quality software. *Process Engineering, 73*(6), 43–46.

Anonymous. (1995). Improving product quality and spotting potential problems with software. *Electrical Design and Manufacturing, 9*(4), 13–14.

Anonymous. (2001). Software, design and quality control. *Modern Plastics, 78*(10), 34–38.

April, A., Abran, A., & Dumke, R. R. (2004). SMCMM model to evaluate and improve the quality of the software maintenance process. *Proceedings of the European Conference on Software Maintenance and Reengineering,* 243–248.

Ares, J., & Pazos, J. (1998). Conceptual modeling: An essential pillar for quality software development. *Knowledge-Based Systems, 11*(2), 87–104.

Arnoult, W. S. (1991). Quality software development through effective project management. *Proceedings of the Project Management Institute Annual Seminar/ Symposium,* 135–138.

Asada, M., & Yan, P. M. (1998). Strengthening software quality assurance. *Hewlett-Packard Journal, 49*(2), 89–97.

Ashlund, S. L., & Horwitz, K. J. (1996). Usability improvements in Lotus CC: Mail for Windows. *Proceedings of the Conference on Human Factors in Computing Systems,* 481–488.

Ashrafi, N. (1998). Decision making framework for software total quality management. *International Journal of Technology Management, 16*(4–6), 532–543.

Ashrafi, N., Berman, O., & Cutler, M. (1994). Optimal design of large software systems using N-version programming. *IEEE Transactions on Reliability, 43,* 344–350.

Avizienis, A. (1967). Design of fault-tolerant computers. *Proceedings of the Conference of the American Federation of Information Processing Societies,* 733–743.

Avizienis, A. (1971). Fault-tolerant computing: An overview. *Computer, 1971*(January/ February), 5–8.

Ayari, N., Barbaron, D., & Lefevre, L. (2008). On improving the reliability of Internet services through active replication. *Proceedings of the International Conference on Parallel and Distributed Computing,* 259–262.

Aykin, N. (1994). Software reuse: A case study on cost-benefit of adopting a common software development tool. In R. G. Bias & D. J. Mayhew (Eds.), *Cost justifying usability* (pp. 177–202). San Francisco, CA: Kaufmann.

Azuma, M., Komiyama, T., & Miyake, T. (1990). Panel: The model and metrics for software quality evaluation report of the Japanese National Working Group. *Proceedings of the 14th Annual International Computer Software and Applications Conference,* 64–69.

Bachiochi, D., Berstene, M., Chouinard, E., Conlan, N., Danchak, M., Furey, T., et al. (1997). Usability studies and designing navigational aids for the World Wide Web. *Computer Networks and ISDN Systems, 29*(8–13), 1489–1496.

Bailey, J. E., & Pearson, S. W. (1983). Development of a tool for measuring and analyzing computer user satisfaction. *Management Science, 29*(6), 519–529.

Baisch, E., & Liedtke, T. (1997). Comparison of conventional approaches and soft-computing approaches for software quality prediction. *Proceedings of the IEEE International Conference on Systems, Man, and Cybernetics,* 1045–1049.

Bajaj, A., & Krishnan, R. (1999). CMU-WEB: A conceptual model for designing usable Web applications. *Journal of Database Management, 10*(4), 33–34.

Baker, R. A. J. (1997). Code reviews enhance software quality. *Proceedings of the IEEE 19th International Conference on Software Engineering,* 570–571.

Ball, M., & Hardie, F. (1969). Majority voter considerations for a TMR computer. *Computer Design, 1969*(April), 100–104.

Barbeau, M. (1999). Implementation of two approaches for the reliable multicast of mobile agents over wireless networks. *Proceedings of the International Symposium on Parallel Architectures, Algorithms and Networks, I-SPAN,* 414–419.

Bargas-Avila, J. A., Oberholzer, G., Schmutz, P., DeVito, M., & Opwis, K. (2007). Usable error message presentation in the World Wide Web: Do not show errors right way. *Interacting with Computers, 19*(3), 330–341.

Barker, R. T., & Biers, D. W. (1994). Software usability testing: Do user self-consciousness and the laboratory environment make any difference? *Proceedings of the Human Factors and Ergonomics Society Conference,* 1131–1134.

Barrett, T. (1997). Dancing with devils: Or facing the music on software quality. *Electronic Design, 45*(13), 117–120.

Beaver, J. M., & Schiavone, G. A. (2003). A comparison of software quality modeling techniques. *Proceedings of the International Conference on Software Engineering Research and Practice,* 263–266.

Beckner, S. A., & Mottay, F. E. (2001). Global perspective on Web site usability. *IEEE Software, 18*(1), 54–61.

Beizer, B. (1984). *Software system testing and quality assurance.* New York, NY: Van Nostrand Reinhold.

Bell, S. J., & Halperin, M. (1995). Testing the reliability of cellular online searching. *Online (Wilton, CT), 19*(5), 15–24.

Benbunan-Fich, R. (2001). Using protocol analysis to evaluate the usability of a commercial Web site. *Information Management, 39*(2), 151–163.

Bennett, K. H. (1997). Software maintenance: A tutorial. In M. Dorfman & R. H. Thayer (Eds.), *Software engineering* (pp. 289–303). Los Alamitos, CA: IEEE Computer Society Press.

Bennett, P. (1996). Experience in engineering quality into software. *Software Engineering Journal, 11*(2), 95–98.

Bevan, N., & Azuma, M. (1997). Quality in use: Incorporating human factors into the software engineering lifecycle. *Proceedings of the 3rd IEEE International Software Engineering Standards Symposium and Forum,* 169–179.

Binder, L. H., & Poore, J. H. (1990). Field experiments with local software quality metrics. *Software—Practice and Experience, 20*(7), 631–647.

Bishop, D. C., & Pymms, P. (1991). Software quality management. *Nuclear Engineer: Journal of the Institution of Nuclear Engineers, 32*(4), 128–131.

Blaha, M. (2004). A copper bullet for software quality improvement. *Computer, 37*(2), 21–25.

Bodsberg, L. (1993). A comparative study of quantitative models for hardware, software, and human reliability assessment. *Quality and Reliability Engineering International, 9,* 501–518.

Boegh, J., Depanfilis, S., & Kitchenham, B. (1999). Method for software quality planning, control, and evaluation. *IEEE Software, 16*(2), 69–77.

Boland, P. J., & Singh, H. (2003). A birth-process approach to Moranda's geometric software reliability model. *IEEE Transactions on Reliability, 52,* 168–174.

Bolchini, D., & Garzotto, F. (2007). Quality Web usability evaluation methods: An empirical study on MILE. *Proceedings of the Web Information Systems Engineering (WISE) International Workshop,* 481–492.

Bolstad, M. (2004). Design by contract: A simple technique for improving the quality of software. *Proceedings of the DoD HPCMP Users Group Conference,* 303–307.

Bonald, T., & Massoulie, L. (2001). Impact of fairness on Internet performance. *Proceedings of the Joint International Conference on Measurement and Modeling of Computer Systems, 29,* 82–91.

Boone, R., Lucas, K., & Wynd, R. (2003). Practical quality metrics for resolution enhancement software. *Proceedings of the SPIE Cost and Performance in Integrated Circuit Creation Conference, 5043,* 162–171.

Borgerson, B. R., & Freitas, R. F. (1975). A reliability model for gracefully degrading and standby-sparing systems. *IEEE Transactions on Computers, C-24,* 517–525.

Bouktif, S., Kegl, B., & Sahraoui, H. (2002). Combining software quality predictive models: An evolutionary approach. *Proceedings of the IEEE International Conference on Software Maintenance,* 385–392.

Bouricius, W. G., Carter, W. C., & Jessep, D. C. (1971). Reliability modeling of fault-tolerant computers. *IEEE Transactions on Computers, C-20,* 1306–1311.

Brandt, W. D., & Wilson, D. N. (1994). Measure of software quality. *Proceedings of the 2nd International Conference on Software Quality Management,* 73–75.

Brooks, F. (1987). No silver bullet: Essence and accidents of software engineering. *IEEE Computer, 20*(4), 10–19.

Brosins, D. B., & Jurison, J. (1973). Design of a voter comparator switch for redundant computer modules. *Proceedings of the IEEE Symposium on Fault-Tolerant Computing,* 113–118.

Bukowski, J. V., & Goble, W. M. (1990). Practical lessons for improving software quality. *Proceedings of the Annual Reliability and Maintainability Symposium,* 436–440.

Bunce, W. L. (1980). Hardware and software: An analytical approach. *Proceedings of the Annual Reliability and Maintainability Symposium,* 209–213.

Bunse, C., Verlage, M., & Giese, P. (1998). Improved software quality through improved development process descriptions. *Automatica, 34*(1), 23–32.

Burns, P. J. (1993). Software metrics: The key to quality software on the NCC project. *Proceedings of the Networks Technology Conference,* 27–30.

Burr, A. (2001). Quality software: A success story? *Quality World, 27*(3), 4–5.

Burton, S., Swanson, K., & Leonard, L. (1993). Quality and knowledge in software engineering. *AI Magazine, 14*(4), 43–50.

Bush, M. (1990). Improving software quality: The use of formal inspections at the Jet Propulsion Laboratory. *Proceedings of the 12th International Conference on Software Engineering,* 196–199.

Butler, K. A., Jacob, R. J. K., & John, B. E. (1995). Introduction and overview to human-computer interaction. *Proceedings of the Conference on Human Factors in Computing Systems*, 345–346.

Calero, C., & Ruiz, J. (2005). Classifying Web metrics using the Web Quality Model. *Online Information Review*, 29(3), 227–248.

Cangussu, J. W., Mathur, A. P., & Karcich, R. M. (2004). Software release control using defect-based quality estimation. *Proceedings of the 15th International Symposium on Software Reliability Engineering*, 440–450.

Card, D. N. (1990). Software quality engineering. *Information and Software Technology*, 32(1), 1–10.

Card, D. N. (2002). Managing software quality with defects. *Proceedings of the 26th Annual International Computer Software and Applications Conference*, 472–474.

Card, S. K., Moran, T. P., & Newell, A. (1983). *The psychology of human-computer interaction*. Hillsdale, NJ: Lawrence Erlbaum Associates.

Carlson, R., Hobby, R., & Newman, H. B. (2003). Measuring end-to-end Internet performance. *Network Magazine*, 18(4), 42–46.

Caro, A., et al. (2008). A proposal for a set attributes relevant for Web portal data quality. *Software Quality Journal*, 16(4), 513–542.

Carroll, J. M. (2000). *Making use: Scenario-based design of human-computer interactions*. Cambridge, MA: MIT Press.

Carroll, J. M., & Rosson, M. B. (1985). Usability specifications as a tool in iterative development. In R. Hartson (Ed.), *Advances in human computer interaction* (pp. 1–28). Norwood, NJ: Ablex.

Carter, W. C., & Bouricius, W. G. (1971). A survey of fault tolerant computer architecture and its evaluation. *Computer*, 4(1), 9–16.

Carvallo, J. P., Franch, X., & Grau, G. (2004). QM: A tool for building software quality models. *Proceedings of the 12th IEEE International Requirements Engineering Conference*, 358–359.

Cechich, A., Piattini M., & Vallecillo, A. (2003). *Component-based software quality: Methods and techniques*. New York, NY: Springer.

Cha, S. S. (1993). Management aspect of software safety. *Proceedings of the Eighth Annual Conference on Computer Assurance*, 35–40.

Chen, D. J., Chen, W. C., & Huang, S. K. (1994). Survey of the influence of programming constructs and mechanisms on software quality. *Journal of Information Science and Engineering*, 10(2), 177–201.

Chen, T. M., & Oh, T. H. (1999). Reliable services in MPLS. *IEEE Communications Magazine*, 37(12), 58–62.

Cheung, R. C. (1980). A user-oriented software reliability model. *IEEE Transactions on Software Engineering*, 6(2), 118–125.

Cho, J., & Lee, S. J. (2003). An evaluation model for software quality improvement. *Proceedings of the International Conference on Software Engineering Research and Practise*, 615–620.

Choi, J., Lee, W., & Kwon, D. (2004). Quality characteristics for the test of mobile game software. *Proceedings of the International Conference on Software Engineering Research and Practice*, 564–570.

Choi, J., Park, S., & Chong, K. (2004). ISO/IEC 9126 quality characteristics considering the application fields and the test phases of embedded software. *Proceedings of the International Conference on Software Engineering Research and Practice*, 628–632.

Chou, C., Peng, H., & Hsieh, Y. (2004). The design and development of an Internet safety curriculum in Taiwan. *Proceedings of the Seventh IASTED International Conference on Computers and Advanced Technology in Education*, 546–551.

Chun, K. C., & Tortorella, M. (2001). Spares-inventory sizing for end-to-end service availability. *Proceedings of the International Symposium on Product Quality and Integrity*, 98–102.

Cin, M. D. (1980). Availability analysis of a fault-tolerant computer system. *IEEE Transactions on Reliability, R-29*, 265–268.

Cobb, R. H., & Mills, H. D. (1990). Engineering software under statistical quality control. *IEEE Software, 7*(6), 45–54.

Comfort, W. T. (1983). A fault-tolerant system architecture for navy applications. *IBM Journal of Research and Development, 27*, 219–236.

Conklin, P. F. (1991). Bringing usability effectively into product development. In M. Rudisill, C. Lewis, P. Polson, & T. McKay (Eds.), *Human-computer interface design: Success stories, emerging methods, real-world context* (pp. 367–385). San Francisco, CA: Morgan Kaufmann.

Constantine, L. L., & Lockwood, L. A. D. (1999). *Software for use: A practical guide to the models and methods of usage-centered design*. Reading, MA: Addison-Wesley.

Conte, T., Vaz, V., da Silva, J., Mendes, E., & Travassos, G. (2008). Process model elicitation and a reading technique for Web usability inspections. *Proceedings of the Web Information Systems Engineering (WISE) International Workshop*, 36–47.

Corry, M. D., Frick, T. W., & Hansen, L. (1997). User-centered design and usability testing of a Web site: An illustrative case study. *Educational Technology, Research, and Development, 45*(4), 65–76.

Cosgriff, P. S. (1994). Quality assurance of medical software. *Journal of Medical Engineering & Technology, 18*(1), 1–10.

Cosgriff, P. S., Vauramo, E., & Pretschner, P. (1992). QAMS: Quality Assurance of Medical Software. *Proceedings of the Advances in Medical Informatics: Results of the AIM Exploratory Action Conference*, 310–313.

Costabile, M. F. (1999). Usable multimedia applications. *Proceedings of the International Conference on Multimedia Computing and Systems*, 124–127.

Costes, A., Landrault, C., & Laprie, J. C. (1978). Reliability and availability models for maintained systems featuring hardware failures and design faults. *IEEE Transactions on Computers, C-27*(6), 400–405.

Cox, M. E., O'Neal, P., & Pendley, W. L. (1994). UPAR analysis: Dollar measurement of a usability indicator for software products. In R. G. Bias & D. J. Mayhew (Eds.), *Cost justifying usability* (pp. 143–158). San Francisco, CA: Morgan Kaufmann.

Cox, S. J., & Tait, N. R. S. (1991). *Reliability, safety, risk management*. London: Butterworth-Heinemann Ltd.

Craig, G. R. (1974). *Software reliability study* (Report No. RADC-TR-74-250). Rome, NY: Rome Air Development Center, Griffiss Air Force Base.

Crovella, M., Lindemann, C., & Reiser, M. (2000). Internet performance modeling: The state of the art at the turn of the century. *Performance Evaluation, 42*(2), 91–108.

Daughtrey, T. (Ed.). (2002). *Fundamental concepts for the software quality engineer*. Milwaukee, WI: ASQ Quality Press.

Davis, C. J., Thompson, J. B., & Smith, P. (1993). Current approaches to software quality assurance within the United Kingdom. *Proceedings of the International Conference on Software Quality Management*, 849–854.

Delcambre, S. N., Rainey, V. P., & Tanik, M. M. (1995). Defining quality to support the software engineering activity. *Proceedings of the Energy-Sources Technology Conference and Exhibition*, 93–99.

Demirors, O. (1997). Assumptions and difficulties of software quality movement. *Proceedings of the 23rd EUROMICRO Conference*, 115–122.

Dette, W. (1990). Software product assurance: IBM'S software quality assurance function. *Forensic Engineering*, 2(1–2), 89–96.

Dhillon, B. S. (1999). *Design reliability*. Boca Raton, FL: CRC Press.

Dhillon, B. S. (2000). *Medical device reliability and associated areas*. Boca Raton, FL: CRC Press.

Dhillon, B. S., & Singh, C. (1981). *Engineering reliability: New techniques and applications*. New York, NY: John Wiley and Sons.

Dickson, J. (1991). The RQMS as an evolving system of software quality measurements. *Proceedings of the International Conference on Communications*, 1743–1746.

DiLucca, G. A., et al. (2002). Testing Web applications. *Proceedings of the International Conference on Software Maintenance*, 310–319.

Divan, D. M., Brumsickle, W. E., & Luckiiff, G. A. (2003). Real-time power quality and reliability monitoring for utilities and their C and I customers. *Proceedings of the Sixth International Conference on Advances in Power System Control, Operation and Management*, 49–53.

Dolle, P., & Jackson, K. (1993). Experiences of software quality management in software maintenance: A case study. *Proceedings of the International Conference on Software Quality Management*, 629–634.

Drake, T. (1996). Measuring software quality: A case study. *Computer*, 29(11), 78–87.

Dromey, R. G. (1995). Model for software product quality. *IEEE Transactions on Software Engineering*, 21(2), 146–162.

Dromey, R. G., Bailes, C., & Xiaoge, L. (1993). Model for enhancing software product quality. *Proceedings of the 16th Australian Computer Science Conference*, 461–465.

Dubrovin, V. I., & Doroshenko, Y. N. (2001). Software quality estimation. *Upravlyayushchie Sistemy i Mashiny, 2001*(5), 34–39.

Dugan, J. B., Sullivan, K. J., & Coppit, D. (1999). Developing a high-quality software tool for fault tree analysis. *Proceedings of the International Symposium on Software Reliability Engineering*, 222–231.

Dumke, R. R., & Kuhran, I. (1994). Tool-based quality management in object-oriented software development. *Proceedings of the 3rd Symposium on Assessment of Quality Software Development Tools*, 148–160.

Duncan, S. P., Martin, C. R., & Quigley-Lawrence, R. (1990). "Customers" and "users": Two faces of software quality and productivity. *Proceedings of the IEEE International Conference on Communications*, 15–18.

Dunn, R., & Ullman, R. (1982). *Quality assurance for computer software*. New York, NY: McGraw Hill.

Dyer, M. (1992). Cleanroom approach to quality software development. *Proceedings of the CMG'92 Conference*, 1201–1212.

Ebert, C., & Morschel, I. (1997). Metrics for quality analysis and improvement of object-oriented software. *Information and Software Technology*, 39(7), 497–509.

Ebert, C., & Morschel, I. (1999). Environment for measuring and improving the quality of object-oriented software. *Quality and Reliability Engineering International*, 15(1), 33–45.

Edwards, A. (1995). *Extraordinary human-computer interactive: Interfaces for users with disabilities.* Cambridge, UK: Cambridge University Press.

Eguchi, T., Ohsaki, H., & Murata, M. (2002). Multivariate analysis for performance evaluation of active queue management mechanisms in the Internet. *Internet Performance and Control of Network Systems, 2002,* 144–153.

Eickelmann, N., & Hayes, J. H. (2004). New year's resolutions for software quality. *IEEE Software, 21*(1), 12–13.

Elboushi, M. I., & Sherif, J. S. (1997). Object-oriented software design utilizing quality function deployment. *Journal of Systems and Software, 38*(2), 133–143.

Elenien, A. R. A., Ismail, L. S., & Bedor, H. S. (2004). Quality of service handler for MPEG video in best effort environment. *Proceedings of the International Conference on Electrical, Electronic and Computer Engineering,* 393–398.

Ericson, C. A. (1981). Software and system safety. *Proceedings of the 5th International System Safety Conference, 1,* III B.1–III B.11.

Erikkson, I., & McFadden, F. (1993). Quality function deployment: A tool to improve software quality. *Information and Software Technology, 35*(9), 491–498.

Esaki, K., Yamada, S., & Takahashi, M. (2002). A quality engineering approach to human factors affecting software reliability in design process. *Electronics and Communications in Japan, 85*(3), 33–42.

Eslambolchi, H., & Danishmand, M. (2003). *Reliability of emerging Internet based services.* New York, NY: Wiley.

Evans, I. (2004). *Achieving software quality through teamwork.* Boston, MA: Artech House.

Faith, B. J. (1993). Making a better job of software quality with Total Quality Management. *Proceedings of the International Conference on Software Quality Management,* 73–76.

Fakhre-Zakeri, I., & Slud, E. (1995). Mixture models for reliability of software with imperfect debugging: Identifiability of parameters. *IEEE Transactions on Reliability, 44,* 104–113.

Fallah, M. H., & Jrad, A. M. (1994). SQA: A proactive approach to assuring software quality. *AT&T Technical Journal, 73*(1), 26–33.

Farber, C. G. (1993). Software installation in a distributed computing environment: Human factors issues and methods. *Proceedings of the Human Factors and Ergonomics Society Conference,* 277–281.

Farbey, B. (1990). Software quality metrics: Considerations about requirements and requirement specifications. *Information and Software Technology, 32*(1), 60–64.

Feng, J., Tang, R., & Wang, S. (2005). Study on software quality grey quantitative evaluation mode. *Harbin Gongye Daxue Xuebao/Journal of Harbin Institute of Technology, 37*(5), 639–641.

Fernandez, A., Insfran, E., & Abrahao, S. (2009). Integrating a usability model into model-driven Web development processes. *Proceedings of the Web Information Systems Engineering International Workshop,* 497–510.

Ferre, Z., Juristo, N., Windl, H., & Constantine, L. (2001). Usability basics for software developers. *IEEE Software, 18*(1), 22–29.

Field, J., & Varela, C. A. (2005). Transactors: A programming model for maintaining globally consistent distributed state in unreliable environments. *ACM SIGPLAN Notices, 40*(1), 195–208.

Fink, R. A. (1998). Reliability modeling of freely available Internet-distributed software. *Proceedings of the International Software Metrics Symposium,* 101–104.

Finnie, B. W., & Johnston, I. H. A. (1990). Acceptance of software quality as a contributor to safety. *Proceedings of the Safety and Reliability Society Symposium*, 279–284.

Fishburn, S. (1992). Integration of a Total Quality Management program through software aided design, qualification, planning, and scheduling tools. *Proceedings of the International SAVE Conference*, 233–239.

Flamm, L. E. (1989). Usability testing: Where does it belong in software design? *Proceedings of the IEEE International Conference on Systems, Man, and Cybernetics*, 235–236.

Fortier, S. C., & Michael, J. B. (1993). A risk-based approach to cost-benefit analysis of software safety activities. *Proceedings of the Eighth Annual Conference on Computer Assurance*, 53–60.

Freeman, H. A., & Metze, G. (1972). Fault-tolerant computers using dotted logical redundancy techniques. *IEEE Transactions on Computers, C-21*, 867–871.

Frey, G. (2002). Software quality in logic controller programming. *Proceedings of the IEEE International Conference on Systems, Man and Cybernetics*, 516–521.

Friedman, M. A., & Voas, J. M. (1995). *Software assessment*. New York, NY: John Wiley and Sons.

Fuccella, J. (1997). Using user centered design methods to create and design usable Web sites. *Proceedings of the Special Interest Group on Design of Communication (SIGDOC) Conference*, 69–77.

Galdi, V., Ippolito, L., & Piccolo, A. (1996). On industrial software quality assessment. *Proceedings of the 3rd International Conference on Software for Electrical Engineering Analysis and Design*, 501–509.

Galin, D. (2004). *Software quality assurance*. New York, NY: Pearson Education Limited.

Ganesan, K., Khoshgoftaar, T. M., & Allen, E. B. (2000). Case-based software quality prediction. *International Journal of Software Engineering and Knowledge Engineering, 10*(2), 139–152.

Ghods, M., & Nelson, K. M. (1998). Contributors to quality during software maintenance. *Decision Support Systems, 23*(4), 361–369.

Gibbs, W. W. (1994). Software's chronic crisis. *Scientific American, 1994*(September), 86–95.

Gillan, D. J., Breedin, S. D., & Cooke, N. J. (1992). Network and multidimensional representations of the declarative knowledge of human-computer interface design experts. *International Journal of Man-Machine Studies, 36*, 587–615.

Glass, R. L. (1976). *Software reliability guidebook*. Rome, NY: Rome Air Development Center, Griffiss Air Force Base.

Godrich, K. L. (2002). Parameterization of powering solutions for telecom/datacom clients. *Proceedings of the 24th International Telecomunications Energy Conference*, 273–278.

Gong, B., Yen, D. C., & Chou, D. C. (1998). Manager's guide to total quality software design. *Industrial Management and Data Systems, 98*(3–4), 100–107.

Gonzalez, R. R. (1995). Unified metric of software complexity: Measuring productivity, quality, and value. *Journal of Systems and Software, 29*(1), 17–37.

Gorton, I., Chan, T. S., & Jelly, I. (1994). Engineering high quality parallel software using PARSE. *Proceedings of the 3rd Joint International Conference on Vector and Parallel Processing*, 381.

Goseva-Popstojanova, K., Mazimdar, S., & Singh, A. D. (2004). Empirical study of session-based workload and reliability for Web servers. *Proceedings of the International Symposium on Software Reliability Engineering*, 403–414.

Goseva-Popstojanova, K., & Trivedi, K. S. (2003). Architecture-based approaches to software reliability prediction. *Computers and Mathematics with Applications, 46*(7), 1023–1036.

Gowen, L. D., & Yap, M. Y. (1993). Traditional software development's effects on safety. *Proceedings of the 6th Annual IEEE Symposium on Computer-Based Medical Systems,* 58–63.

Granic, A., & Glavinic, U. (2002). Usability evaluation issues for computerized educational systems. *Proceedings of the Mediterranean Electrotechnical Conference,* 558–562.

Greco, R. (1997). Satellite: Boosting Internet performance. *Telecommunications (International Edition), 31*(4), 4–9.

Greenfield, D. (2004). Storage, heal thyself! *Network Magazine, 19*(7), 52–53.

Guglielmi, N., & Guerrieri, R. (1994). Experimental comparison of software methodologies for image based quality control. *Proceedings of the 20th International Conference on Industrial Electronics, Control and Instrumentation,* 1942–1945.

Gulezian, R. (1995). Software quality measurement and modeling, maturity, control and improvement. *Proceedings of the 2nd IEEE International Software Engineering Standards Symposium,* 52–59.

Gyorkos, J., Rozman, I., & Vajde Horvat, R. (1996). Quality management in software development process: An empirical model. *Proceedings of the 1996 IEEE International Engineering Management Conference,* 191–195.

Haag, S., Raja, M. K., & Schkade, L. L. (1996). Quality function deployment usage in software development. *Communications of the ACM, 39*(1), 41–49.

Hadfield, P., & Hope, S. (1993). Survey of classification schemes for improving software quality and an outline for their future. *Proceedings of the International Conference on Software Quality Management,* 613–617.

Haitani, L. N. (1992). Applying manufacturing quality control theory to performance-oriented software development. *Proceedings of the CMG '92 Conference,* 890–894.

Hall, T., & Wilson, D. (1997). Views of software quality: A field report. *Proceedings of the IEE Software Conference, 144*(2), 111–118.

Hammer, W., & Price, D. (2001). *Occupational safety management and engineering.* Upper Saddle River, NJ: Prentice Hall.

Hanna, M. S., & Loew-Blosser, W. (1992). SQEngineer: A methodology and tool for specifying and engineering software quality. *Proceedings of the Symposium on Assessment of Quality Software Development Tools,* 194–210.

Hansen, M. D. (1989). Survey of available software-safety analysis techniques. *Proceedings of the Annual Reliability and Maintainability Symposium,* 46–49.

Haque, T. (1997). PCM: The smart way to achieve software quality. *Quality World, 23*(11), 922–923.

Haugh, J. M. (1991). Never make the same mistake twice: Using configuration control and error analysis to improve software quality. *Proceedings of the IEEE/AIAA 10th Digital Avionics Systems Conference,* 220–225.

Haungs, M., Pandey, R., & Barr, E. (2004). Handling catastrophic failures in scalable Internet applications. *Proceedings of the International Symposium on Applications and the Internet,* 188–194.

He, Z., Staples, G., & Ross, M. (1996). Fourteen Japanese quality tools in software process improvement. *TQM Magazine, 8*(4), 40–44.

Head, A. J. (1999). Web redemption and the promise of usability. *Online, 23*(6), 20–32.

Hecht, H. (1972). Economics of reliability for spacecraft computers. *Proceedings of the Reliability and Maintainability Symposium*, 554–564.

Hecht, H., & Hecht, M. (1986). Software reliability in the system context. *IEEE Transactions on Software Engineering*, 12(1), 51–58.

Hecht, M. (2001). Reliability/availability modeling, prediction, and measurement for e-commerce and other Internet information systems. *Proceedings of the Annual Reliability and Maintainability Symposium*, 176–182.

Henry, S., & Goff, R. (1991). Comparison of a graphical and a textual design language using software quality metrics. *Journal of Systems and Software*, 14(3), 133–146.

Herbsleb, J., Zubrow, D., & Goldenson, D. (1997). Software quality and the capability maturity model. *Communications of the ACM*, 40(6), 30–40.

Herron, D. (1998). Early life cycle identification of software quality risk factors. *Proceedings of the ASQ's 52nd Annual Quality Congress*, 399–400.

Hevner, A. R. (1997). Phase containment metrics for software quality improvement. *Information and Software Technology*, 39(13), 867–877.

Hindel, B. (1993). How to ensure software quality for real time systems. *Control Engineering Practice*, 1(1), 35–41.

Hirayama, M., Sato, H., & Yamada, A. (1990). Practice of quality modeling and measurement on software life-cycle. *Proceedings of the 12th International Conference on Software Engineering*, 98–107.

Hoffmann, H. (1993). How to improve overall system quality by a sensible design of man-machine interaction: Views of a software engineer. *Proceedings of the European Safety & Reliability Conference*, 939–944.

Holcomb, R., & Tharp, A. L. (1991). What users say about software usability. *International Journal of Human-Computer Interaction*, 3(1), 49–78.

Hon, S. E. I. (1990). Assuring software quality through measurements: A buyer's perspective. *Journal of Systems and Software*, 13(2), 117–130.

Hong, G. Y., & Goh, T. N. (2003). Six sigma in software quality. *TQM Magazine*, 15(6), 364–373.

Hopcroft, J. E., & Krafft, D. B. (1987). Sizing the U.S. software industry. *IEEE Spectrum*, 1987(December), 58–62.

Hopker, O. (1995). Pragmatic approach to software quality for information systems in small- and medium-sized organisations. *Proceedings of the 3rd International Conference on Software Quality Management*, 75–80.

Horch, J. W. (2003). *Practical guide to software quality management* (2nd ed.). Boston, MA: Artech House.

Horgan, J. R., London, S., & Lyu, M. R. (1994). Achieving software quality with testing coverage measures. *Computer*, 27(9), 60–70.

Houston, D., & Keats, J. B. (1998). Cost of software quality: A means of promoting software process improvement. *Quality Engineering*, 10(3), 563–567.

Howard, S., Hammond, J., & Lindgaard, G. (Eds.). (1997). *Human-computer interaction: Interact'97*. London: Chapman and Hall.

Hrasnica, H., & Lehnert, R. (2002). Performance analysis of error handling methods applied to a broadband PLC access network. *Proceedings of the Internet Performance and Control of Network Systems Conference*, 4865, 166–177.

Hsiao, M. Y., Carter, W. C., Thomas, J. W., & Stringfellow, W. R. (1981). Reliability, availability, and serviceability of IBM computer systems: A quarter century of progress. *IBM Journal of Research and Development*, 25, 453–465.

Hudson, L. (2000). Radical usability (or, Why you need to stop redesigning your Web site). *Library Computing, 19*(1–2), 86–92.

Hun Oh, S., Yeon Lee, J., & Lee, Y. J. (1995). Software Quality Manager: A knowledge-based management tool of software metrics. *Proceedings of the 1994 IEEE Region 10's 9th Annual International Conference (TENCON'94),* 796–800.

Huo, M., Verner, J., & Zhu, L. (2004). Software quality and agile methods. *Proceedings of the 28th Annual International Computer Software and Applications Conference,* 520–525.

IEEE 1228-1994. (1994, May). *IEEE standard for software safety plans* (IEEE Std. 1228-1994). New York, NY: Institute of Electrical and Electronic Engineers (IEEE).

Ippolito, L. M., & Wallace, D. R. (1995, January). *A study on hazard analysis in high integrity software standards and guidelines* (Report No. NISTIR 5589). Washington, DC: National Institute of Standards and Technology, U.S. Department of Commerce.

Jack, R. (1994). Customer involvement in the application of quality management systems to software product development. *Proceedings of the Computing and Control Division Colloquium on Customer Driven Quality in Product Design,* 6.1–6.5.

Jacko, J. A., Sears, A., & Sorensen, S. J. (2001). Framework for usability: Healthcare professionals and the Internet. *Ergonomics, 44*(11), 989–1007.

Jacob, R. J. K. (1983). Using formal specifications in the design of a human-computer interface. *Communications of the ACM, 26*(4), 259–264.

Jauw, J., & Vassiliou, P. (2000). Field data is reliability information: Implementing an automated data acquisition and analysis system. *Proceedings of the Annual Reliability and Maintainability Symposium,* 86–93.

Jeffery, R., Basili, V., & Berry, M. (1994). Establishing measurement for software quality improvement. *Proceedings of the IFIP TC8 Open Conference on Business Process Re-engineering: Information Systems Opportunities and Challenges,* 319–329.

Jegers, K. (2008). Investigating the applicability of usability heuristics for evaluation of pervasive games. *Proceedings of the 3rd International Conference on Internet and Web Applications and Services,* 656–661.

Johnson, C. S., & Wilson, D. N. (1995). Using quality principles to teach software quality techniques. *International Journal of Environmental Studies A & B, 47*(1), 465–468.

Johnson, P. M. (1994). Instrumented approach to improving software quality through formal technical review. *Proceedings of the 16th International Conference on Software Engineering,* 113–122.

Jones, D. R., Murthy, V., & Blanchard, J. (1992). Quality and reliability assessment of hardware and software during the total product life cycle. *Quality and Reliability Engineering International, 8*(5), 477–483.

Jones, W. D., Hudepohl, J. P., & Khoshgoftaar, T. M. (1999). Application of a usage profile in software quality models. *Proceedings of the 3rd European Conference on Software Maintenance and Reengineering,* 148–157.

Jorgensen, M. (1997). Measurement of software quality. *Proceedings of the 1st International Conference on Software Quality Engineering,* 257–266.

Jorgensen, M. (1999). Software quality measurement. *Advances in Engineering Software, 30*(12), 907–912.

Joshi, S. M., & Misra, K. B. (1991). Quantitative analysis of software quality during the design and implementation phase. *Microelectronics and Reliability, 31*(5), 879–884.

Joyce, E. (1987). Software bugs: A matter of life and liability. *Datamation, 33*(10), 88–92.

Juliff, P. (1994). Software quality function deployment. *Proceedings of the 2nd International Conference on Software Quality Management*, 533–536.

Jung, H., & Choi, B. (1999). Optimization models for quality and cost of modular software systems. *European Journal of Operational Research, 112*(3), 613–619.

Jung, H., Kim, S., & Chung, C. (2004). Measuring software product quality: A survey of ISO/IEC 9126. *IEEE Software, 21*(5), 88–92.

Kan, S. H. (2003). *Metrics and models in software quality engineering.* Boston, MA: Addison-Wesley.

Kan, S. H., Basili, V. R., & Shapiro, L. N. (1994). Software quality: An overview from the perspective of Total Quality Management. *IBM Systems Journal, 33*(1), 4–19.

Kan, S. H., Dull, S. D., & Amundson, D. N. (1994). AS/400 software quality management. *IBM Systems Journal, 33*(1), 62–88.

Kantner, L., & Rosenbaum, S. (1997). Usability studies of WWW sites: Heuristic evaluation versus laboratory testing. *Proceedings of the Special Internet Group on Design of Communication (SIGDOC) Conference*, 153–160.

Kaplan, C. (1996). Secrets of software quality at IBM. *Proceedings of the ASQC's 50th Annual Quality Congress*, 479–484.

Karat, C. (1993). Usability engineering in dollars and cents. *IEEE Software, 10*(3), 88–89.

Karat, C. (1997). Cost-justifying usability. In M. G. Helander, T. K. Landauer, & P. V., Prabhu (Eds.), *The handbook of human-computer interaction* (pp. 767–781). Amsterdam: North-Holland.

Karat, J., & Dayton, T. (1995). Practical education for improving software usability. *Proceedings of the Conference on Human Factors in Computing Systems*, 162–169.

Kaufer, S., & Stacy, W. (1996). Software quality begins with a quality design. *Object Magazine, 5*(8), 56–60.

Kautz, K., & Ramzan, F. (2001). Software quality management and software process improvement in Denmark. *Proceedings of the 34th Annual Hawaii International Conference on System Sciences*, 126–129.

Keene, S. J. (1991). Cost-effective software quality. *Proceedings of the Annual Reliability and Maintainability Symposium*, 433–437.

Keene, S. J. (1992). Assuring software safety. *Proceedings of the Annual Reliability and Maintainability Symposium*, 274–279.

Keene, S. J. (1992). Software reliability concepts. *Annual Reliability and Maintainability Symposium Tutorial Notes, (1992)*, 1–21.

Keene, S., & Lane, C. (1992). Combined hardware and software aspects of reliability. *Quality and Reliability Engineering International, 8*(5), 419–426.

Kellogg, W. A., & Breen, T. J. (1987). Evaluating user and system models: Applying scaling techniques to problems in human-computer interaction. *Proceedings of the Conference on Graphics Interface*, 303–309.

Kelly, D. P., & Oshana, R. S. (2000). Improving software quality using statistical testing techniques. *Information and Software Technology, 42*(12), 801–807.

Kemp, A. (1995). Software quality giving managers what they need. *Engineering Management Journal, 5*(5), 235–238.

Kenett, R. S. (1996). Software specifications metrics: A quantitative approach to assess the quality of documents. *Proceedings of the 19th Convention of Electrical and Electronics Engineers in Israel*, 166–169.

Kermarrec, A., Massoulie, L., & Ganesh, A. J. (2003). Probabilistic reliable dissemination in large-scale systems. *IEEE Transactions on Parallel and Distributed Systems, 14*(3), 248–258.

Khaled, E. E. (2005). *The ROI from software quality.* Boca Raton, FL: Auerbach Publications.

Khoshgoftaar, T. M., & Allen, E. B. (1997). Impact of costs of misclassification on software quality modeling. *Proceedings of the 4th International Software Metrics Symposium,* 54–62.

Khoshgoftaar, T. M., & Allen, E. B. (2000). A practical classification rule for software quality models. *IEEE Transactions on Reliability, 49*(2), 209–216.

Khoshgoftaar, T. M., Allen, E. B., & Deng, J. (2001). Controlling overfitting in software quality models: Experiments with regression trees and classification. *Proceedings of the 7th International Software Metrics Symposium,* 190–198.

Khoshgoftaar, T. M., Allen, E. B., & Halstead, R. (1998). Using process history to predict software quality. *Computer, 31*(4), 66–72.

Khoshgoftaar, T. M., Allen, E. B., & Kalaichelvan, K. S. (1996). Predictive modeling of software quality for very large telecommunications systems. *Proceedings of the IEEE International Conference on Communications,* 214–219.

Khoshgoftaar, T. M., Halstead, R., & Allen, E. B. (1997). Process measures for predicting software quality. *Proceedings of the High-Assurance Systems Engineering Workshop,* 155–160.

Khoshgoftaar, T. M., & Lanning, D. L. (1994). On the impact of software product dissimilarity on software quality models. *Proceedings of the 4th International Symposium on Software Reliability Engineering,* 104–114.

Khoshgoftaar, T. M., & Munson, J. C. (1992). Software metrics and the quality of telecommunication software. *Proceedings of Tricomm: High-Speed Communications Networks,* 255–260.

Khoshgoftaar, T. M., Munson, J. C., & Bhattacharya, B. B. (1992). Predictive modeling techniques of software quality from software measures. *IEEE Transactions on Software Engineering, 18*(11), 979–987.

Khoshgoftaar, T. M., & Seliya, N. (2003). Fault prediction modeling for software quality estimation: Comparing commonly used techniques. *Empirical Software Engineering, 8*(3), 255–283.

Khoshgoftaar, T. M., & Seliya, N. (2004). Comparative assessment of software quality classification techniques: An empirical case study. *Empirical Software Engineering, 9*(3), 229–257.

Khoshgoftaar, T. M., Seliya, N., & Herzberg, A. (2005). Resource-oriented software quality classification models. *Journal of Systems and Software, 76*(2), 111–126.

Khoshgoftaar, T. M., Shan, R., & Allen, E. B. (2000). Improving tree-based models of software quality with principal components analysis. *Proceedings of the 11th International Symposium on Software Reliability Engineering,* 198–209.

Khoshgoftaar, T. M., Szabo, R. M., & Guasti, P. J. (1995). Exploring the behaviour of neural network software quality models. *Software Engineering Journal, 10*(3), 89–96.

Khoshgoftaar, T. M., Yuan, X., & Allen, E. B. (2000). Balancing misclassification rates in classification-tree models of software quality. *Empirical Software Engineering, 5*(4), 313–330.

Kikuchi, N., Mizuno, O., & Kikuno, T. (2000). Identifying key attributes of projects that affect the field quality of communication software. *Proceedings of the IEEE 24th Annual International Computer Software and Applications Conference,* 176–178.

Kim, D., Hong, W., & Jong, M. (2001). A fault management system for reliable ADSL services provisioning. *Proceedings of the Internet Performance and Control of Network Systems Conference, 4523,* 341–349.

Kim, J. R., & Kim, D. (2008). Enhancing Internet network reliability with integrated framework of multi-objective genetic algorithm and Monte Carlo simulation. *Asia-Pacific Journal of Operational Research, 25*(6), 837–846.

Kim, K. H., & Welch, H. O. (1989). Distributed execution of recovery approach for uniform treatment of hardware and software faults applications. *IEEE Transactions on Computers, 38*, 626–636.

Kirakowski, J., & Cierlik, B. (1998). Measuring the usability of Web sites. *Proceedings of the Human Factors Society Conference*, 424–428.

Kitchenham, B., & Pfleeger, S. L. (1996). Software quality: The elusive target. *IEEE Software, 13*(1), 12–21.

Kline, M. B. (1980). Software and hardware reliability and maintainability: What are the differences? *Proceedings of the Annual Reliability and Maintainability Symposium*, 179–185.

Knox, S. T. (1993). Modeling the cost of software quality. *Digital Technical Journal, 5*(4), 9–12.

Koczela, L. J. (1971). A three-failure-tolerant computer system. *IEEE Transactions on Computers, 1971*(November), 1389–1393.

Koethluk, F. A. (1995). EURESCOM project P227. Quality assurance of software for telecommunication systems. *Proceedings of the 3rd International Conference on Software Quality Management*, 51–55.

Kogan Y., & Choudhury, G. (2004). Two problems in Internet reliability: New questions for old models. *ACM Sigmetrics Performance Evaluation Review, 32*(2), 9–11.

Kokol, P., Chrysostalis, M., & Bogonikolos, N. (2003). Software quality founded on design laws. *Proceedings of the Seventh IASTED International Conference on Software Engineering and Applications*, 728–732.

Kopetz, H. (1979). *Software reliability*. London: Macmillan.

Koval, D. O., & Ewasechko, H. F. (1985). Digital computer systems reliability. *Proceedings of the Annual Reliability and Maintainability Symposium*, 69–76.

Krasner, H. (1999). Exploring the principles that underlie software quality costs. *Proceedings of the ASQ Annual Quality Congress*, 500–503.

Kuehn, R. E. (1969). Computer redundancy: Design, performance and future. *IEEE Transactions on Reliability, R-18*, 3–11.

Lac, C., & Raffy, J. (1992). Tool for software quality. *Proceedings of the Symposium on Assessment of Quality Software Development Tools*, 144–150.

Lai, L. T., Yun, X., & Tan, F. B. (2009). Attributes of Web site usability: A study of Web users with the repertory grid technique. *International Journal of Electronic Commerce, 13*(4), 97–126.

Laitenberger, O. (1998). Studying the effects of code inspection and structural testing on software quality. *Proceedings of the 9th International Symposium on Software Reliability Engineering*, 237–246.

Landauer, T. K. (1995). *The trouble with computers: Usefulness, usability, and productivity*. Cambridge, MA: MIT Press.

Lanubile, F., & Visaggio, G. (1997). Evaluating predictive quality models derived from software measures: Lessons learned. *Journal of Systems and Software, 38*(3), 225–234.

Lauesen, S. (1997). Usability engineering in industrial practice. In S. Howard, J. Hammond, & G. Lindgaard (Eds.), *Human-computer interaction* (pp. 15–22). London: Chapman and Hall.

Lauesen, S., & Younessi, H. (1998). Is software quality visible in the code? *IEEE Software, 15*(4), 69–73.

Lazar, J., Meiselwitz, G., & Norcio, A. (2004). A taxonomy of novice user perception of error on the Web. *Universal Access in the Information Society, 3*(3–4), 202–208.

Le Gall, G. (1995). Software quality control. *Commutation & Transmission, 17*(3), 91–101.

Lecerof, A., & Paterno, F. (1998). Automatic support for usability evaluation. *IEEE Transactions on Software Engineering, 24*(10), 863–888.

Lee, M., Shih, K., & Huang, T. (2001). Well-evaluated cohesion metrics for software quality. *Ruan Jian Xue Bao/Journal of Software, 12*(10), 1447–1463.

Lee, S. (2001). The reliability improvement of TCP for the wireless packet data network by using the virtual TCP receive window. *Proceedings of the IEEE Vehicular Technology Conference*, 1866–1868.

Lee, S., Kim, J., & Choi, J. (2004). Development of a Web-based integrity evaluation system for primary components in a nuclear power plant. *Proceedings of the Asian Pacific Conference on Nondestructive Testing*, 2226–2231.

Leffingwell, D. A., & Norman, B. (1993). Software quality in medical devices: A top-down approach. *Proceedings of the 6th Annual IEEE Symposium on Computer-Based Medical Systems*, 307–311.

Lehner, F. (1993). Quality control in software documentation: Measurement of text comprehensibility. *Information & Management, 25*(3), 133–137.

Lelouche, R., & Page, M. (2001). Save on time and effort: Application of discount usability engineering to the design of Internet-based learning environments. *International Journal of Continuing Engineering Education, 11*(1–2), 35–46.

Lem, E. (2005). *Software metrics: The discipline of software quality.* North Charleston, SC: Book Surge.

Leveson, N. G. (1984). Software safety in computer-controlled systems. *IEEE Computer, 17*(2), 48–55.

Leveson, N. G. (1986). Software safety: Why, what, and how. *Computing Surveys, 18*(2), 125–163.

Leveson, N. G. (1995). *Safeware.* Reading, MA: Addison-Wesley.

Leveson, N. G., & Harvey, P. R. (1983). Analyzing software safety. *IEEE Transactions on Software Engineering, 9*(5), 569–579.

Leveson, N. G., & Shimeall, T. J. (1991). Safety verification of ADA programs using software fault trees. *IEEE Software, 8*(4), 48–59.

Levine, B. N., Lavo, D. B., & Garcia-Luna-Aceves, J. J. (1996). Case for reliable concurrent multicasting using shared ACK trees. *Proceedings of the ACM International Multimedia Conference*, 365–376.

Lew, P., Olsina, L., & Li, Z. (2010). Quality, quality in use, actual usability and user experience as key drivers for Web application evaluation. *Proceedings of the International Conference on Web Engineering*, 218–232.

Lewis, J. R. (1995). IBM computer usability satisfaction questionnaires: Psychometric evaluation and instructions for use. *International Journal of Human-Computer Interaction, 7*(1), 57–78.

Lewis, N. D. C. (1999). Assessing the evidence from the use of SPC in monitoring, predicting and improving software quality. *Computers and Industrial Engineering, 37*(1–2), 157–160.

Li, F. (2008). Usability evaluation on websites. *Proceedings of the 9th International Conference on Computer-Aided Industrial Design and Conceptual Design: Multicultural Creation and Design*, 242–244.

Li, J. J., & Hogan, J. R. (1998). To maintain a reliable software specification. *Proceedings of the International Symposium on Software Reliability Engineering*, 59–68.

Lim, A. (1997). Advanced techniques for maintaining reliability of complex computer systems. *Proceedings of the Hawaii International Conference on System Sciences*, 4–13.

Lin, H. X., Choong, Y. Y., & Salvendy, G. (1997). Proposed index of usability: A method for comparing the relative usability of different software systems. *Behavior and Information Technology*, 16(4–5), 267–278.

Lin, W. W. K., & Wong, A. K. Y. (2003). A novel adaptive fuzzy logic controller (FLC) to improve Internet channel reliability and response timeliness. *Proceedings of the 8th IEEE Symposium on Computers and Communications*, 1347–1352.

Linberg, K. R. (1993). Defining the role of software quality assurance in a medical device company. *Proceedings of the 6th Annual IEEE Symposium on Computer-Based Medical Systems*, 278–283.

Lindermeier, R. (1994). Quality assessment of software prototypes, *Reliability Engineering & System Safety*, 43(1), 87–94.

Lipow, M. (1979). Prediction of software failures. *J. Syst. Software*, 1, 71–75.

Lippy, B. E. (2001). Technology Safety Data Sheets: The U.S. Department of Energy's innovative efforts to mitigate or eliminate hazards during design and to inform workers about the risks of new technologies. *Proceedings of the International Conference on Radioactive Waste Management and Environmental Remediation*, 375–379.

Littlewood, B., Popov, P., & Strigini, L. (2002). Assessing the reliability of diverse fault-tolerant software-based systems. *Safety Science*, 40, 781–796.

Liu, H., & Shooman, M. L. (2003). Reliability computation of an IP/ATM network with congestion. *Proceedings of the International Symposium on Product Quality and Integrity: Transforming Technologies for Reliability and Maintainability Engineering*, 581–586.

Liu, X. F. (1998). Quantitative approach for assessing the priorities of software quality requirements. *Journal of Engineering and Applied Science*, 42(2), 105–113.

Liu, Z., Almhana, J., & Choulakian, V. (2004). Internet performance modeling using mixture dynamical system models. *Proceedings of the International Conference on Pervasive Services*, 189–198.

Lloyd, I. J., & Simpson, M. J. (1993). Legal aspects of software quality. *International Conference on Software Quality Management*, 247–252.

Lombardi, F. (1982). Availability modelling of ring microcomputer systems. *Microelectronic and Reliability*, 22, 295–308.

Long, D., Caroll, J. L., & Park, C. J. (1991). A study of the reliability of Internet sites. *Proceedings of the 10th Symposium on Reliable Distributed Systems*, 177–186.

Long, D., Muir, A., & Golding, R. (1995). A longitudinal survey of Internet host reliability. *Proceedings of the 14th Symposium on Reliable Distributed Systems*, 2–9.

Longbottom, L. (1972). Analysis of computer systems reliability and maintainability. *The Radio and Electronic Engineer*, 42(December), 110–114.

Longbottom, R. (1980). *Computer system reliability*. Chichester, UK: John Wiley and Sons.

Lounis, H., & Ait-Mehedine, L. (2004). Machine-learning techniques for software product quality assessment. *Proceedings of the Fourth International Conference on Quality Software*, 102–109.

Lowe, J., Daughtrey, T., & Jensen, B. (1993). Software quality: International perspectives. *Proceedings of the 47th Annual Quality Congress*, 893–894.

Lowe, J. E., & Jensen, B. (1992). Customer service approach to software quality. *Proceedings of the 46th Annual Quality Congress*, 1077–1083.

Lowry, E. S. (2004). Software simplicity and hence safety: Thwarted for decades. *Proceedings of the International Symposium on Technology and Society,* 80–84.

Luo, X., Zhan, J., & Mao, M. (2005). New model of improving quality management in China Software Company. *Journal of Computational Information Systems, 1*(1), 175–177.

Lyons, K., & Starner, T. (2001). Mobile capture for wearable computer usability testing. *Proceedings of the International Symposium on Wearable Computers,* 69–76.

Lyu, M. R. (Ed.). (1996). *Handbook of software reliability engineering.* New York, NY: McGraw Hill.

MacIntyre, F., Estep, K. W., & Sieburth, J. M. (1990). Cost of user-friendly programming. *Journal of Forth Application and Research, 6*(2), 103–115.

Macmichael, R. A. (2001). Seven factors to consider when redesigning your site. *IT Professional, 3*(4), 35–37.

Maison, F. P. (1971). The MECRA: A self-repairable computer for highly reliable process. *IEEE Transactions on Computers, C-20,* 1382–1393.

Mallory, S. R. (1993). Building quality into medical product software design. *Biomedical Instrumentation & Technology, 27*(2), 117–135.

Mander, R., & Smith, B. (2002). *Web usability for dummies.* New York, NY: Hungry Minds.

Mandeville, W. A. (1990). Software costs of quality. *IEEE Journal on Selected Areas in Communications, 8*(2), 315–318.

Mantei, M. M., & Theorey, T. J. (1988). Cost/benefit analysis for incorporating human factors in the software life cycle. *Communications of the ACM, 31*(4), 428–439.

Mantyla, M. V. (2004). Developing new approaches for software design quality improvement based on subjective evaluations. *Proceedings of the 26th International Conference on Software Engineering,* 48–50.

Mao, C. Y., & Lu, Y. S. (2004). Testing and evaluation for Web usability based on extended mark chain model. *Wuhan University Journal of Natural Sciences, 9,* 687–693.

Mao, M., & Luo, X. (2004). Software quality management and software process model. *Journal of Information and Computation Science, 1*(3), 203–207.

Marriott, P. C., McCorkhill, B. S., & Lamb, R. A. (1990). Software quality networking for professionals. *Proceedings of the Forty-Fourth Annual Quality Congress Transactions,* 511–517.

Maselko, W. T., Winstead, L. S., & Kahn, R. E. (1999). Streamlined software development process: A proven solution to the cost/schedule/quality challenge. *Proceedings of the Annual Forum of the American Helicopter Society, 1,* 12–35.

Mathur, F. P. (1969). On reliability modelling and analysis of a dynamic TMR system utilizing standby spares. *Proceedings of the Seventh Annual Allerton Conference on Circuits and Systems,* 115–120.

Matthews, W., & Cottrell, L. (2000). PingER Project: Active Internet performance monitoring for the HENP community. *IEEE Communications Magazine, 38*(5), 130–136.

McDonough, J. A. (1990). Template for software quality management for Department of Defense programs. *Proceedings of the IEEE National Aerospace and Electronics Conference,* 1281–1283.

McDowell, A., Schmidt, C., & Yue, K. (2004). Analysis and netrics of XML schema. *Proceedings of the International Conference on Software Engineering Research and Practice,* 538–544.

McParland, P. (1995). Using function point analysis to help assess the quality of a software system. *Proceedings of the 3rd International Conference on Software Quality Management*, 127–130.

Mellor, E. (1987). Experiments in software reliability estimation. *Reliability Engineering*, 18, 117–129.

Mendis, K. S. (1999). Software safety and its relation to software quality assurance. In G. G. Schulmeyer & J. I. McManus (Eds.), *Handbook of software quality assurance* (pp. 669–679). Upper Saddle River, NJ: Prentice Hall.

Merryman, P. M., & Avizienis, A. A. (1975). Modeling transient faults in TMR computer system. *Proceedings of the Annual Reliability and Maintainability Symposium*, 333–339.

Meskens, N. (1996). Software quality analysis systems: A new approach. *Proceedings of the IEEE 22nd International Conference on Industrial Electronics, Control, and Instrumentation*, 1406–1411.

Methodology for software reliability prediction and assessment. (1992). (Report No. RL-TR-92-52, Vols. I and II). Rome, NY: Rome Air Development Center, Griffiss Air Force Base.

Meulen, M. V. D. (2000). *Definitions for hardware and software safety engineers.* London: Springer-Verlag.

Miller, S., & Jarrett, C. (2001). Setting usability requirements for a Web site containing a form. *Proceedings of the Society for Technical Communication Annual Conference*, 386–390.

MIL-STD-498. (1994). *Software development and documentation.* Washington, DC: Department of Defense.

MIL-STD-882D. (2000). *System safety program requirements.* Washington, DC: Department of Defense.

Mirel, B., Olsen, L. A., & Prakash, A. (1997). Improving quality in software engineering through emphasis on communication. *Proceedings of the ASEE Annual Conference*, 9–12.

Moeller, E. W. (2001). The latest Web trend: Usability? *Proceedings of the IEEE International Professional Communication Conference*, 151–158.

Moeller, E. W. (2002). Usability: A Web trend for the future? *Proceedings of the Society for Technical Communication Annual Conference*, 291–296.

Moh, M., & Zhang, S. (2002). Scalability study of application-level reliable multicast congestion control for the next-generation mobile Internet. *Proceedings of the International Conference on Third Generation Wireless and Beyond*, 652–657.

Mohageg, M., Myers, R., Marrin, C., Kent, J., Mott, D., & Isaacs, P. (1996). A user interface for accessing 3D content on the World Wide Web. *Proceedings of the Conference on Human Factors in Computing Systems*, 466–472.

Mohamed, W. E. A., & Siakas, K. V. (1995). Assessing software quality management maturity (SQMM): A new model incorporating technical as well as cultural factors. *Proceedings of the 3rd International Conference on Software Quality Management*, 325–328.

Molich, R., & Nielsen, J. (1990). Improving a human-computer dialogue. *Communications of the ACM*, 33(3), 338–348.

Molina, F., & Toval, A. (2009). Integrating usability requirements that can be evaluated in design time into model driven engineering of Web information systems. *Advances in Engineering Software*, 40(12), 1306–1317.

Monk, A., Wright, P., Haber, J., & Davenport, L. (1992). *Improving your human-computer interface: A practical approach.* Englewood Cliffs, NJ: Prentice-Hall.

Moore, B. J. (1994). Achieving software quality through requirements analysis. *Proceedings of the IEEE International Engineering Management Conference*, 78–83.

Moores, T. T., & Champion, R. E. M. (1994). Software quality through the traceability of requirements specifications. *Proceedings of the 1st International Conference on Software Testing, Reliability and Quality Assurance*, 100–104.

Morgan, M. R. P. (1995). Crossing disciplines: Usability as a bridge between system, software, and documentation. *Technical Communication, 42*(2), 303–306.

Morris, M. G., & Dillon, A. P. (1996). Importance of usability in the establishment of organizational software standards for end user computing. *International Journal of Human-Computer Studies, 45*(2), 243–258.

Morris, M. G., & Turner, J. M. (2001). Assessing users' subjective quality of experience with World Wide Web: An exploratory examination of temporal changes in technology acceptance. *International Journal of Human-Computer Studies, 54*(6), 877–901.

Moses, J. (1993). Software quality methodologies. *Proceedings of the International Conference on Software Quality Management*, 333–337.

Moxey, C. (1993). Experiences with quality on a recent software development project. *Proceedings of the International Conference on Software Quality Management*, 375–378.

Munson, J. C., & Khoshgoftaar, T. M. (1990). Regression modelling of software quality: Empirical investigation. *Information and Software Technology, 32*(2), 106–114.

Murine, G. E. (1995). Using the Rome Laboratory Framework and Implementation Guidebook as the basis for an international software quality metric standard. *Proceedings of the 2nd IEEE International Software Engineering Standards Symposium*, 61–70.

Murugesan, S. (1994). Attitude towards testing: A key contributor to software quality. *Proceedings of the 1st International Conference on Software Testing, Reliability and Quality Assurance*, 111–115.

Musa, J. D. (1975). A theory of software reliability and its applications. *IEEE Transactions on Software Engineering, 1*, 312–317.

Musa, J. D., Iannino, A., & Okumoto, K. (1987). *Software reliability*. New York, NY: McGraw Hill.

Myers, G. J. (1976). *Software reliability: Principles and practices*. New York, NY: John Wiley and Sons.

Myers, G. J. (1979). *The art of software testing*. New York, NY: John Wiley and Sons.

Nagarajan, S. V., Garcia, O., & Croll, P. (2003). Software quality issues in extreme programming. *Proceedings of the 21st IASTED International Multi-Conference on Applied Informatics*, 1090–1095.

Nance, R. E. (2002). *Managing software quality: A measurement framework for assessment and prediction*. New York, NY: Springer.

Nathan, P. (2005). *System testing with an attitude: An approach that nurtures front-loaded software quality*. New York, NY: Dorset House.

Neal, M. L. (1991). Managing software quality through defect trend analysis. *Proceedings of the PMI Annual Seminar/Symposium*, 119–122.

Nelson, E. (1978). Estimating software reliability from test data. *Microelectronics and Reliability, 17*, 67–75.

Neufelder, A. (1993). *Ensuring software reliability*. New York, NY: Marcel Dekker.

Nielsen, J. (1989). What do users really want? *International Journal of Human-Computer Interaction, 1*(2), 137–147.

Nielsen, J. (1992). The usability engineering life cycle. *Computer, 25,* 12–22.

Nielsen, J. (1992). Finding usability problems through heuristic evaluation. *Proceedings of the ACM Conference on Human Factors in Computing Systems,* 373–380.

Nielsen, J. (1993). Iterative user-interface design. *Computer, 26*(11), 32–41.

Nielsen, J. (1995). Getting usability used. In K. Nordby, P. Helmersen, D. J. Gilmore, & S. Arnesen (Eds.), *Human-computer interaction* (pp. 3–12). London: Chapman and Hall.

Nielsen, J. (1995). Usability inspection methods. *Proceedings of the Conference on Human Factors in Computing Systems,* 377–378.

Nielsen, J. (1996). Usability metrics: Tracking interface improvements. *IEEE Software, 13,* 12–13.

Nielsen, J. (2000). *Designing Web usability: The practice of simplicity.* Indianapolis, IN: New Riders.

Nielsen, J., & Faber, J. M. (1996). Improving system usability through parallel design. *Computer, 29*(2), 29–35.

Nielsen, J., & Landauser, T. K. (1993). Mathematical model of the finding of usability problems. *Proceedings of the Conference on Human Factors in Computing Systems,* 206–213.

Nielsen, J., & Levy, J. (1994). Measuring usability preference vs. performance. *Communications of the ACM, 37*(4), 66–75.

Nielsen, J., & Sano, D. (1995). Sun Web: User interface design for Sun Microsystem's internal Web. *Computer Networks and ISDN Systems, 28*(1–2), 179–188.

Norman, D. A. (1983). Design rules based on analyses of human error. *Communications of the ACM, 26*(4), 254–258.

Offutt, J. (2002). Quality attributes of Web software applications. *IEEE Software, 19*(2), 25–32.

Ogasawara, H., Yamada, A., & Kojo, M. (1996). Experiences of software quality management using metrics through the life-cycle. *Proceedings of the 18th International Conference on Software Engineering,* 179–188.

Ogawa, K. (1992). Evaluation method of computer usability based on human-to-computer information transmission model, *Ergonomics, 35*(5–6), 557–590.

Olagunju, A. O. (1992). Concepts of operational software quality metrics. *Proceedings of the 20th Annual ACM Computer Science Conference,* 301–308.

Opiyo, E. Z., Horvath, I., & Vergeest, J. S. M. (2002). Quality assurance of design support software: Review and analysis of the state of the art. *Computers in Industry, 49*(2), 195–215.

O'Regan, G. (2002). *A practical approach to software quality.* New York, NY: Springer.

Osaki, S., & Nishio, T. (1980). *Reliability evaluation of some fault-tolerant computer architectures.* Berlin: Springer Verlag.

Osman, T., Wagealla, W., & Bargiela, A. (2004). An approach to rollback recovery of collaborating mobile agents. *IEEE Transactions on Systems, Man and Cybernetics Part C: Applications and Reviews, 34*(1), 48–57.

Osmundson, J. S., Michael, J. B., & Machniak, M. J. (2003). Quality management metrics for software development. *Information and Management, 40*(8), 799–812.

Oztekin, A. (2011). A decision support system for usability evaluation of Web-based information systems. *Expert Systems with Applications, 38*(3), 2110–2118.

Paolini, P. (1999). Hypermedia, the Web and usability issues. *Proceedings of the International Conference on Multimedia Computing & Systems,* 111–115.

Parnas, D. L., & Lawford, M. (2003). Inspection's role in software quality assurance. *IEEE Software, 20*(4), 16–20.

Parnas, D. L., & Lawford, M. (2003). The role of inspection in software quality assurance. *IEEE Transactions on Software Engineering, 29*(8), 674–676.

Parzinger, M. J., & Nath, R. (1997). Effects of TQM implementation factors on software quality. *Proceedings of the Annual Meeting of the Decision Sciences Institute*, 834–836.

Patel, R. B., & Mastorakis, N. (2005). Fault-tolerant mobile agents computing. *WSEAS Transactions on Computers, 4*(3), 287–314.

Patki, V. B., Patki, A. B., & Chatterji, B. N. (1983). Reliability and maintainability considerations in computer performance evaluation. *IEEE Transactions on Reliability, R-32*, 433–436.

Paulish, D. J. (1990). Methods and metrics for developing high quality patient monitoring system software. *Proceedings of the Third Annual IEEE Symposium on Computer-Based Medical Systems Conference*, 145–152.

Pawlicki, G., & Sathaye, A. (2004). Availability and performance oriented availability modeling of webserver cache hierarchies. *Proceedings of the Annual Reliability and Maintainability Symposium: International Symposium on Product Quality and Integrity*, 586–592.

Pearrow, M. (2000). *Web site usability handbook*. Rockland, MA: Charles River Media.

Pedrycz, W., Peters, J. F., & Ramanna, S. (1998). Software quality measurement: Concepts and fuzzy neural relational model. *Proceedings of the IEEE International Conference on Fuzzy Systems*, 1026–1031.

Pelnik, T. M., & Suddarth, G. J. (1998). Implementing training programs for software quality assurance engineers. *Medical Device and Diagnostic Industry, 20*(10), 75–80.

Pence, J. L., & Hon, S. E. (1990). Software surveillance: A buyer quality assurance program. *IEEE Journal on Selected Areas in Communications, 8*(2), 301–308.

Peslak, A. R. (2004). Improving software quality: An ethics-based approach. *Proceedings of the 2004 ACM SIGMIS CPR Conference*, 144–149.

Pfleeger, S. L. (1998). Software quality. *Dr. Dobb's Journal of Software Tools for Professional Programmer, 23*(3), 22–27.

Phan, D. D., George, J. F., & Vogel, D. R. (1995). Managing software quality in a very large development project. *Information & Management, 29*(5), 277–283.

Pieper, M., & Hermsdorf, D. (1997). BSCW for disabled teleworkers: Usability evaluation and interface adaptation of an Internet-based cooperation environment. *Computer Network and ISDN Systems, 29*(8–13), 1479–1487.

Pierce, K. M. (1994). Why have a software quality assurance program? *Nuclear Plant Journal, 12*(5), 57–61.

Pivka, M. (1995). Software quality system in a small software house. *Proceedings of the 3rd International Conference on Software Quality Management*, 83–86.

Plant, R. T. (1991). Factors in software quality for knowledge-based systems. *Information and Software Technology, 33*(7), 527–536.

Post, D. E., & Kendall, R. P. (2004). Software project management and quality engineering practices for complex, coupled multiphysics, massively parallel computational simulations: Lessons learned from ASCI. *International Journal of High Performance Computing Applications, 18*(4), 399–416.

Potter, P. (1989). Usability labs make a difference. *Insurance Software Review, 14*(6), 15–17.

Pradhan, D. K. (Ed.). (1986). *Fault-tolerant computing theory and techniques* (Vols. 1 and 2). Englewood Cliffs, NJ: Prentice-Hall.

Pratt, W. M. (1991). Experiences in the application of customer-based metrics in improving software service quality. *Proceedings of the International Conference on Communications*, 1459–1462.

Putnik, Z. (1998). On the quality of the software produced. *Proceedings of the International Conference on Computer Science and Informatics*, 64–66.

Rai, A., Song, H., & Troutt, M. (1998). Software quality assurance: An analytical survey and research prioritization. *Journal of Systems and Software, 40*(1), 67–83.

Ramakrishnan, S. (1994). Quality factors for resource allocation problems: Linking domain analysis and object-oriented software engineering. *Proceedings of the 1st International Conference on Software Testing, Reliability and Quality Assurance*, 68–72.

Ramakumar, R. (1993). *Engineering reliability: Fundamentals and applications*. Englewood Cliffs, NJ: Prentice Hall.

Ramamoorthy, C. V. (1971). Fault tolerant computing: An introduction and an overview. *IEEE Transactions on Computers, C-20*, 1241–1244.

Ramamoorthy, C. V. (2002). Evolution and evaluation of software quality models. *Proceedings of the 14th International Conference on Tools with Artificial Intelligence*, 543–546.

Ramamoorthy, C. V., & Bastani, F. B. (1982). Software reliability: Status and perspectives. *IEEE Transactions on Software Engineering, 8*, 354–371.

Raman, S. (1997). CMM: A road map to software quality. *Proceedings of the 51st Annual Quality Congress*, 898–906.

Raman, S., & McCanne, S. (1999). Model analysis and protocol framework for soft state-based communication. *Computer Communication Review, 29*(4), 15–25.

Ramanna, S. (2002). Approximation methods in a software quality measurement framework. *Proceedings of the IEEE Canadian Conference on Electrical and Computer Engineering*, 566–571.

Ramanna, S., Peters, J. F., & Ahn, T. (2002). Software quality knowledge discovery: A rough set approach. *Proceedings of the 26th Annual International Computer Software and Applications Conference*, 1140–1145.

Ravden, S. J., & Johnson, G. I. (1989). *Evaluating usability of human-computer interfaces: A practical method*. Chichester, UK: Ellis Horwood.

Redig, G., & Swanson, M. (1990). Control Data Corporation's government systems group standard software quality program. *Proceedings of the IEEE National Aerospace and Electronics Conference*, 670–674.

Redig, G., & Swanson, M. (1993). Total Quality Management for software development. *Proceedings of the 6th Annual IEEE Symposium on Computer-Based Medical Systems*, 301–306.

Redmill, F. J. (1990). Considering quality in the management of software-based development projects. *Information and Software Technology, 32*(1), 18–25.

Rideout, T. B., Uyeda, K. M., & Williams, E. L. (1989). Evolving the software usability engineering process at Hewlett-Packard. *Proceedings of the IEEE International Conference on Systems, Man, and Cybernetics*, 229–234.

Rigby, P. J., Stoddart, A. G., & Norris, M. T. (1990). Assuring quality in software: Practical experiences in attaining ISO 9001. *British Telecommunications Engineering, 8*(4), 244–249.

Riley, C. A., & McConkie, A. B. (1989). Designing for usability: Human factors in a large software development organization. *Proceedings of the IEEE International Conference on System, Man, and Cybernetics*, 225–228.

Robertson, J. W. (1994). Usability and children's software: A user-centered design methodology. *Journal of Computing in Childhood Education, 5*(3/4), 257–271.

Rodford, J. (2000). New levels of software quality by automatic software inspection. *Electronic Engineering (London), 72*(882), 3–9.

Roland, H. E., & Moriarty, B. (1983). *System safety engineering and management.* New York, NY: John Wiley and Sons.

Rosenberg, L. H., & Sheppard, S. B. (1994). Metrics in software process assessment, quality assurance and risk assessment. *Proceedings of the 2nd International Software Metrics Symposium*, 10–16.

Rozum, J. A. (1997). Roadmap to improving software productivity and quality. *Proceedings of the ISA TECH/EXPO Technology Update Conference*, 185–194.

Rupe, J. (2003). Performability modeling and decision support for computer telephony integration. *Proceedings of the International Symposium on Product Quality and Integrity: Transforming Technologies for Reliability and Maintainability Engineering*, 339–343.

Russell, B., & Chatterjee, S. (2003). Relationship quality: The undervalued dimension of software quality. *Communications of the ACM, 46*(8), 85–89.

Sackman, H., Erikson, W. J., & Grant, E. E. (1968). Exploratory experimentation studies comparing online and offline programming performance. *Communications of the ACM, 11,* 3–11.

Sahinoglu, M., & Libby, D. L. (2003). Sahinoglu-Libby (SL) probability density function-component reliability applications in integrated networks. *Proceedings of the International Symposium on Product Quality and Integrity: Transforming Technologies for Reliability and Maintainbility Engineering*, 280–287.

Sakasai, Y., & Hotta, K. (1996). Construction of software quality assurance system. *NTT R&D, 45*(3), 237–246.

Salameh, W. A. (1994). Comparative study on software quality assurance using complexity metrics. *Advances in Modelling and Analysis A, 22*(1), 1–6.

Salvador, P., & Valadas, R. (2001). A framework based on Markov-modulated Poisson processes for modeling traffic with long-range dependence. *Proceedings of the Internet Performance and Control of Network Systems Conference*, 221–232.

Sanso, B., & Mellah, H. (2009). On reliability, performance and Internet power consumption. *Proceedings of the 7th International Workshop on Design of Reliable Communication Networks*, 259–264.

Saxena, A., & Bhatia, S. (2005). Identity Management for improved organizational efficiency and e-business growth: Managing enterprise knowledge. *International Journal of Information Technology and Management, 4*(3), 321–342.

Scerbo, M. W. (1991). Usability engineering approach to software quality. *Proceedings of the 45th Annual Quality Congress*, 726–733.

Scerbo, M. W. (1994). Usability engineering approach to software quality. *Annual Quality Congress Transactions, 45,* 726–733.

Schick, G. J., & Wolverton, R. W. (1978). An analysis of competing software reliability models. *IEEE Transactions on Software Engineering, 4,* 104–120.

Schilling, W. W., & Alam, M. (2007). Measuring the reliability of existing Web servers. *Proceedings of the IEEE International Conference on Electro/Information Technology*, 399–404.

Schneider, P., & Hines, M. L. A. (1990). Classification of medical software. *Proceedings of the IEEE Symposium on Applied Computing*, 20–27.

Schneidewind, N. F. (1993). Report on the IEEE standard for a software quality metrics methodology. *Proceedings of the Conference on Software Maintenance*, 104–106.

Schneidewind, N. F. (1997). Software metrics model for integrating quality control and prediction. *Proceedings of the 8th International Symposium on Software Reliability Engineering*, 402–415.

Schneidewind, N. F. (1997). Software metrics model for quality control. *Proceedings of the 4th International Software Metrics Symposium*, 127–136.

Schneidewind, N. F. (1997). Reliability modelling for safety-critical software. *IEEE Transactions on Reliability*, 46(1), 88–98.

Schneidewind, N. F. (1999). Software quality maintenance model. *Proceedings of the Conference on Software Maintenance*, 277–286.

Schneidewind, N. F. (2001). Knowledge requirements for software quality measurement. *Empirical Software Engineering*, 6(3), 201–205.

Schneidewind, N. F. (2002). Body of knowledge for software quality measurement. *Computer*, 35(2), 77–83.

Scholtz, J., & Shneiderman, B. (1999). Introduction to special issue on usability engineering. *Empirical Software Engineering*, 4(1), 5–10.

Schoonmaker, S. J. (1992). Engineering software quality management. *Proceedings of the Energy-Sources Technology Conference and Exhibition*, 55–63.

Schultz, E. E., et al. (2001). Usability and security: An appraisal of usability issues in information security methods. *Computers and Security*, 20(7), 620–634.

Schwefel, H., Jobmann, M., & Hollisch, D. (2001). On the accuracy of TCP performance models. *Proceedings of the Internet Performance and Control of Network Systems Conference*, 91–102.

Scicchitano, P. (1995). Great ISO 9000 debate: Quality in software. *Compliance Engineering*, 12(7), 4–5.

Scott, R. K., Gault, J. W., & McAllister, D. G. (1987). Fault-tolerant software reliability modeling. *IEEE Transactions on Software Engineering*, 13(5), 106–111.

Sears, A., & Jacko, J. A. (2001). Understanding the relation between network quality of service and the usability of distributed multimedia documents. *Human-Computer Interaction*, 15(1), 43–68.

Sedigh-Ali, S., Ghafoor, A., & Paul, R. A. (2001). Metrics-guided quality management for component-based software systems. *Proceedings of the 25th Annual International Computer Software and Applications Conference*, 303–308.

Sellers, F. F., Hsiao, M., & Bearnson, L. W. (1968). *Error detecting logic for digital computers*. New York, NY: McGraw Hill.

Sengupta, S., & Oliaro, G. (2002). The NSTX trouble reporting system. *Proceedings of the Symposium on Fusion Engineering*, 242–244.

Shackel, B. (1986). IBM makes usability as important as functionality. *The Computer Journal*, 29(3), 10–14.

Shaw, R. (1994). Safety-critical software and current standards initiative. *Computer Methods and Programs in Biomedicine*, 44, 5–22.

Shaw, R. D., & Livingston, R. D. (2002). Web-based occupational health, safety and environmental (OHSE) management tools: Can they help companies manage OHSE performance in a cost-effective manner? *Proceedings of the SPE International Conference on Health, Safety and Environment in Oil and Gas Exploration and Production*, 1551–1554.

Shelden, S., & Vaughan, M. (2001). The Internet usability engineer. *Ergonomics in Design, 9,* 27–28.

Shepperd, M. (1990). Early life-cycle metrics and software quality models. *Information and Software Technology, 32*(4), 311–316.

Sherif, Y. S. (1992). Software safety analysis: The characteristics of efficient technical walkthroughs. *Microelectronics and Reliability, 32*(3), 407–414.

Sherif, Y. S., & Kelly, J. C. (1992). Improving software quality through formal inspections. *Microelectronics and Reliability, 32*(3), 423–431.

Sherif, Y. S., & Kheir, N. A. (1984). Reliability and failure analysis of computer systems. *Computers and Electrical Engineering, 11,* 151–157.

Shi, X., Wang, H., & Zhong, Q. (1998). Design and implementation of a software quality evaluation system. *Jisuanji Gongcheng/Computer Engineering, 24*(2), 8–11.

Shida, T., Yoshinari, Y., & Hisatsune, M. (2001). Development of information technology in the construction and maintenance of nuclear power plants. *Hitachi Review, 50*(3), 73–78.

Shooman, M. L. (1975). Software reliability measurement and models. *Proceedings of the Annual Reliability and Maintainability Symposium,* 485–491.

Simmons, R. A. (1990). Software quality assurance (SQA) early in the acquisition process. *Proceedings of the IEEE National Aerospace and Electronics Conference,* 664–669.

Slaughter, S. A., Harter, D. F., & Krishna, M. S. (1998). Evaluating the cost of software quality. *IEEE Engineering Management Review, 26*(4), 32–37.

Smith, P. A., Newman, I. A., & Parks, L. M. (1997). Virtual hierarchies and virtual networks: Some lessons from hypermedia usability research applied to the World Wide Web. *International Journal of Human-Computer Studies, 47*(1), 67–95.

Smith, S. L., & Mosier, J. N. (1986). *Design guidelines for designing user interface software* (Report No. MTR-10090). Bedford, MA: MITRE Corp.

Soi, I. M., & Gopal, K. (1980). Hardware vs. software reliability: A comparative study. *Microelectronics and Reliability, 20,* 881–885.

Sova, D. W., & Smidts, C. (1996). Increasing testing productivity and software quality: A comparison of software testing methodologies within nasa. *Empirical Software Engineering, 1*(2), 165–188.

Spencer, R. (2000). The streamlined cognitive walkthrough method: Working around social constraints encountered in a software development company. *Proceedings of the Conference on Human Factors in Computing Systems,* 353–359.

Spencer, R. H. (1985). *Computer usability testing and evaluation.* Englewood Cliffs, NJ: Prentice Hall.

Spinelli, A., Pina, D., & Salvaneschi, P. (1995). Quality measurement of software products: An experience about a large automation system. *Proceedings of the 2nd Symposium on Software Quality Techniques and Acquisition Criteria,* 192–194.

Squires, D., & Preece, J. (1999). Predicting quality in educational software: Evaluating for learning, usability and the synergy between them. *Interacting with Computers, 11*(5), 467–483.

Staknis, M. E. (1990). Software quality assurance through prototyping and automated testing. *Information and Software Technology, 32*(1), 26–33.

Stangel, M., & Bharghavan, V. (1998). Improving TCP performance in mobile computing environments. *Proceedings of the IEEE International Conference on Communications,* 584–589.

Stanyer, D., & Procter, R. (1999). Improving Web usability with the Link Lens. *Computer Networks, 31*(11), 1533–1544.

Stavrianidis, P. (1992). Reliability and uncertainty analysis of hardware failures of a programmable electronic system. *Reliability Engineering and System Safety, 39,* 309–324.

Stieber, H. A. (1997). Statistical quality control: How to detect unreliable software components. *Proceedings of the 8th International Symposium on Software Reliability Engineering,* 8–12.

Stockman, S. (1993). Total quality management in software development. *Proceedings of the IEEE Global Telecommunications Conference,* 498–504.

Stockman, S. G., Todd, A. R., & Robinson, G. A. (1990). Framework for software quality measurement. *IEEE Journal on Selected Areas in Communications, 8*(2), 224–233.

Su, S. Y. H., & DuCasse, E. (1980). Hardware redundancy reconfiguration scheme for tolerating multiple module failures. *IEEE Transactions on Computers, 29,* 254–258.

Sukert, A. N. (1977). An investigation of software reliability models. *Proceedings of the Annual Reliability and Maintainability Symposium,* 478–484.

Sulaiman, S. (1996). Usability and the software production life cycle. *Proceedings of the Conference on Human Factors in Computing Systems,* 61–62.

Sullivan, K., & Horwitz, K. J. (1996). Windows 95 user interface: A case study in usability engineering. *Proceedings of the Conference on Human Factors in Computing Systems,* 473–480.

Sullivan, P. (1989). Usability in the computer industry: What contribution can longitudinal field studies make? *Proceedings of the International Conference on Professional Communication Conference (IPCC'89),* 12–16.

Suri, D. (2004). Software quality assurance for software engineers. *Proceedings of the ASEE Annual Conference and Exposition,* 12727–12735.

Suryn, W., Abran, A., & April, A. (2003). ISO/IEC SQuaRE: The second generation of standards for software product quality. *Proceedings of the Seventh IASTED International Conference on Software Engineering and Applications,* 807–814.

Svitek, M. (2003). Architecture of the transport telematics applications using the GNSS. *Proceedings of the International Conference on Information and Knowledge Engineering,* 505–508.

Takagi, H., Kitajima, M., & Yamamoto, T. (2001). Search process evaluation for a hierarchical menu system by Markov chains. *Proceedings of the Internet Performance and Control of Network Systems Conference,* 183–192.

Takahashi, R. (1997). Software quality classification model based on McCabe's complexity measure. *Journal of Systems and Software, 38*(1), 61–69.

Takahashi, R., Muraoka, Y., & Nakamura, Y. (1997). Building software quality classification trees: Approach, experimentation, evaluation. *Proceedings of the 8th International Symposium on Software Reliability Engineering,* 222–233.

Takesue, T. (2000). CG technologies for supporting cooperative creativity by industrial designers. *Proceedings of the IEEE International Workshop on Robot and Human Interactive Communication,* 316–321.

Talbert, N. B. (1993). Representative sampling within software quality assurance. *Proceedings of the Conference on Software Maintenance,* 174–179.

Tanaka, T., Aizawa, M., & Ogasawara, H. (1998). Software quality analysis and measurement service activity in the company. *Proceedings of the International Conference on Software Engineering,* 426–429.

Terazzi, A., Giordano, A., & Minco, G. (1998). How can usability measurement affect the reengineering process of clinical software procedures? *International Journal of Medical Informatics, 52*(1–3), 229–234.

Tervonen, I. (1994). Necessary skills for a software quality engineer. *Proceedings of the 2nd International Conference on Software Quality Management*, 573–577.

Tervonen, I., & Kerola, P. (1998). Towards deeper co-understanding of software quality. *Information and Software Technology, 39*(14–15), 995–1003.

Thayer, R. H. (1997). Software engineering project management. In M. Dorfman & R. H. Thayer (Eds.), *Software engineering* (pp. 358–371). Los Alamitos, CA: IEEE Computer Society Press.

Thayer, T. A., Lipow, M., & Nelson, E. C. (1978). *Software reliability*. New York, NY: North Holland.

Thompson, J. B., & Edwards, H. M. (1994). StePS: A method that will improve software quality. *Proceedings of the 2nd International Conference on Software Quality Management*, 131–144.

Tian, J. (1998). Early measurement and improvement of software quality. *Proceedings of the IEEE 22nd Annual International Computer Software and Applications Conference*, 196–201.

Tian, J. (2005). *Software quality engineering: Testing, quality assurance, and quantifiable improvement*. New York, NY: Wiley.

Tiwari, A., & Tandon, A. (1994). Shaping software quality: The quantitative way. *Proceedings of the 1st International Conference on Software Testing, Reliability and Quality Assurance*, 84–94.

To, M., & Neusy, P. (1994). Unavailability analysis of long-haul networks. *IEEE Journal on Selected Areas in Communications, 12*(1), 100–109.

Todd, A. (1990). Measurement of software quality improvement. *Proceedings of the Colloquium on Software Metrics*, 8.1–8.5.

Toyama, M., Sugawara, M., & Nakamura, K. (1990). High-quality software development system: AYUMI. *IEEE Journal on Selected Areas in Communications, 8*(2), 201–209.

Trammell, C. J., & Poore, J. H. (1992). Group process for defining local software quality: Field applications and validation experiments. *Software: Practice and Experience, 22*(8), 603–636.

Tripathi, P., Pandey, M., & Bharti, D. (2010). Towards the identification of usability metric for academic Web-sites. *Proceedings of the 2nd International Conference on Computer and Automation Engineering*, 393–397.

Tse, P. W., & He, L. S. (2001). Web and virtual instrument based machine remote sensing, monitoring and fault diagnostic system. *Proceedings of the Biennial Conference on Mechanical Vibration and Noise*, 2919–2926.

Tsiaousis, A. S., & Giaglis, G. M. (2010). An empirical assessment of environment factors that influence the usability of a mobile website. *Proceedings of the 9th International Conference on Mobile Business*, 161–167.

Ueyama, Y., & Ludwig, C. (1994). Joint customer development process and its impact on software quality. *IEEE Journal on Selected Areas in Communications, 12*(2), 265–270.

UL. (1998). *Software in programmable components*. Northbrook, IL: Underwriters Laboratories.

Van Waes, L. (2000). Thinking aloud as a method for testing the usability of websites! The influence of task variation on the evaluation of hypertext. *IEEE Transactions on Professional Communication, 43*(3), 279–291.

Veitch, P. (2003). A survivable and cost-effective IP metro interconnect architecture. *IEEE Communications Magazine, 41*(12), 100–105.

Virzi, R. A. (1997). Usability inspection methods. In M. Helander, T. K. Landauer, & P. Prabhu (Eds.), *The handbook of human-computer interaction* (pp. 231–252). Amsterdam: Elsevier Science.

Visaggio, G. (1997). Structural information as a quality metric in software systems organization. *Proceedings of the International Conference on Software Maintenance,* 92–99.

Voas, J. (2002). The COTS software quality challenge. *Proceedings of the 56th Annual Quality Congress,* 93–96.

Voas, J. (2003). Assuring software quality assurance. *IEEE Software, 20*(3), 48–49.

Voas, J., & Agresti, W. W. (2004). Software quality from a behavioral perspective. *IT Professional, 6*(4), 46–50.

Vollman, T. (1994). Standards support for software tool quality assessment. *Proceedings of the 3rd Symposium on Assessment of Quality Software Development Tools,* 29–39.

Von Hellens, L. A. (1997). Information systems quality versus software quality: A discussion from a managerial, an organizational and an engineering viewpoint. *Information and Software Technology, 39*(12), 801–808.

Vora, P. (1995). Classifying user errors in human computer interactive tasks. *Common Ground (Usability Professional Association), 5*(2), 15–16.

Wackerbarth, G. (1990). Quality assurance for software. *Forensic Engineering, 2*(1–2), 97–111.

Waite, D. A. (1994). Software quality management from the outside in. *Proceedings of the ASQC Annual Quality Congress,* 778–782.

Waite, D. A. (1998). Internet information sources: Quantity vs. quality. *Proceedings of the ASQ's 52nd Annual Quality Congress,* 344–348.

Wakerly, J. F. (1976). Reliability of microcomputer systems using triple modular redundancy. *Proceedings of the IEEE Computer Society International Conference,* 23–26.

Waldegg, P. B., & Scrivener, S. A. R. (1996). Designing interfaces for culturally diverse users. *Proceedings of the 6th Australian Conference on Computer-Human Interaction,* 316–317.

Walker, A. J. (1997). Quality management applied to the development of a national checklist for ISO 9001 audits for software. *Proceedings of the 3rd IEEE International Software Engineering Standards Symposium and Forum,* 6–14.

Walker, A. J. (1998). Improving the quality of ISO 9001 audits in the field of software. *Information and Software Technology, 40*(14), 865–869.

Wallace, D. R., Kuhn, D. R., & Ippolito, L. M. (1992). An analysis of selected software safety standards. *Proceedings of the Seventh Annual Conference on Computer Assurance,* 123–136.

Wallmueller, E. (1991). Software quality management. *Microprocessing and Microprogramming, 32*(1–5), 609–616.

Walsh, J. (1994). Software quality management: A subjective view. *Engineering Management Journal, 4*(3), 105–111.

Walton, T. (2001). Quality planning for software development. *Proceedings of the 25th Annual International Computer Software and Applications Conference,* 104–109.

Wang, G., Cao, J., & Chan, K. C. C. (2004). RGB: A scalable and reliable group membership protocol in mobile Internet. *Proceedings of the International Conference on Parallel Processing,* 326–333.

Wattanapongskorn, N., & Coit, D. W. (2007). Fault-tolerant embedded system design and optimization considering reliability estimation uncertainty. *Reliability Engineering and System Safety, 92,* 395–407.

Wegner, E. (1995). Quality of software packages: The forthcoming international standard. *Computer Standards & Interfaces, 17*(1), 115–120.

Wells, C. H., Brand, R., & Markosian, L. (1995). Customized tools for software quality assurance and reengineering. *Proceedings of the 2nd Working Conference on Reverse Engineering*, 71–77.

Welzel, D., & Hausen, H. (1993). Five-step method for metric-based software evaluation: Effective software metrication with respect to quality standards. *Microprocessing and Microprogramming, 39*(2–5), 273–276.

Werth, L. H. (1993). Quality assessment on a software engineering project. *IEEE Transactions on Education, 36*(1), 181–183.

Wesenberg, D. P., & Vansaun, K. (1991). A system approach for software quality assurance. *Proceedings of the IEEE National Aerospace and Electronics Conference*, 771–776.

Wesselius, J., & Ververs, F. (1990). Some elementary questions on software quality control. *Software Engineering Journal, 5*(6), 319–330.

Weyuker, E. J. (1999). Evaluation techniques for improving the quality of very large software systems in a cost-effective way. *Journal of Systems and Software, 47*(2), 97–103.

Wheeler, S., & Duggins, S. (1998). Improving software quality. *Proceedings of the 36th Annual Southeast Conference*, 300–309.

Whiteside, J., Bennett, J., & Holzblatt, K. (1988). Usability engineering: Our experience and evolution. In M. Halander & K. Holzblatt (Eds.), *Handbook of human-computer interaction* (pp. 791–817). New York, NY: Elsevier Science.

Whittaker, J. A., & Voas, J. M. (2002). 50 years of software: Key principles for quality. *IT Professional, 4*(6), 28–35.

Wilcox, R. H. (1962). *Redundancy techniques for computing systems.* New York, NY: Spartan Books.

Wilson, D. (1993). Software quality assurance in Australia. *Proceedings of the International Conference on Software Quality Management*, 911–914.

Wishart, J. (2004). Internet safety in emerging educational contexts. *Computers and Education, 43*(1–2), 193–204.

Wong, B. (2003). Measurements used in Software Quality Evaluation, *Proceedings of the International Conference on Software Engineering Research and Practise*, 971–977.

Woodman, M. (1993). Making software quality assurance a hidden agenda? *Proceedings of the International Conference on Software Quality Management*, 301–305.

Wu, C., Lin, J., & Yu, L. (2005). Research of military software quality characteristic and design attribute. *Jisuanji Gongcheng/Computer Engineering, 31*(12), 100–102.

Wunnava, S. V., & Jasani, H. (2002). Secure multimedia activity with redundant schemes. *Proceedings of the IEEE Southeast Conference*, 333–337.

Xenos, M., & Christodoulakis, D. (1994). Applicable methodology to automate software quality measurements. *Proceedings of the International Conference on Software Testing, Reliability and Quality Assurance*, 121–125.

Xiao-Ying, S., & Lan, Y. (2001). A software quality evaluation system: JT-SQE. *Wuhan University Journal of Natural Sciences, 6*(1–2), 511–515.

Xie, M., Dai, Y. S., & Poh, K. L. (2004). *Computing systems reliability: Models and analysis.* New York, NY: Springer.

Xu, Z., & Khoshgoftaar, T. M. (2001). Software quality prediction for high-assurance network telecommunications systems. *Computer Journal, 44*(6), 558–568.

Yamada, S. (1991). Software quality/reliability measurement and assessment: Software reliability growth models and data analysis. *Journal of Information Processing*, 14(3), 254–266.

Yamaguchi, M., & Yamamoto, M. (2005). Congestion control scheme for reliable multicast improving intrasession fairness with network support. *Electronics and Communications in Japan, Part I: Communications* (English translation of *Denshi Tsushin Gakkai Ronbunshi*), 88(2), 61–70.

Yan, H., Hu, J., & Zhang, L. (2001). Process management for software quality assurance based on document status. *Beijing Hangkong Hangtian Daxue Xuebao/Journal of Beijing University of Aeronautics and Astronautics*, 27(4), 474–477.

Yang, S., & Chou, H. (2003). Adaptive QoS parameters approach to modeling Internet performance. *International Journal of Network Management*, 13(1), 69–82.

Yang, Y. (2000). Synthetic evaluation method for software quality. *Xiaoxing Weixing Jisuanji Xitong/Mini-Micro Systems*, 21(3), 313–315.

Yang, Y. H. (2001). Software quality management and ISO 9000 implementation. *Industrial Management and Data Systems*, 101(7), 329–338.

Yao, B., & Fuchs, W. K. (2000). Proxy-based recovery for applications on wireless hand-held devices. *Proceedings of the IEEE Symposium on Reliable Distributed Systems*, 2–10.

Yau, S. S., Wang, Y., & Huang, J. G. (1990). An integrated expert system framework for software quality assurance. *Proceedings of the 14th Annual International Computer Software and Applications Conference*, 161–166.

Yokoyama, Y., & Kodaira, M. (1998). Software cost and quality analysis by statistical approaches. *Proceedings of the International Conference on Software Engineering*, 465–467.

Yoon, W., Lee, D., & Youn, H. Y. (2002). A combined group/tree approach for scalable many-to-many reliable multicast. *IEEE Infocom*, 3, 1336–1345.

Yoshimura, M., Hamaguchi, N., & Kozaki, T. (2004). High-performance and data-optimized 3rd-generation mobile communication system: 1xEV-DO. *Hitachi Review*, 53(6), 271–275.

Zainudin, N. M., Ahmad, W. F. W., & Goh, K. N. (2010). Evaluating C2C e-commerce website usability in Malaysia from users' perspective: A case study. *Proceedings of the International Symposium on Information Technology*, 151–156.

Zamojski, W., & Caban, D. (2006). Impact of software and failures on the reliability of a man-computer system. *International Journal of Reliability, Quality and Safety Engineering*, 13(2), 149–156.

Zargari, A., Cantrell, P. A., & Grise, W. (1997). Application of statistical process control (SPC) software in Total Quality Improvement (TQI). *Proceedings of the 23rd IEEE Electrical Electronics Insulation Conference and Electrical Manufacturing and Coil Winding*, 829–834.

Zeng, X., Tsai, J. J. P., & Weigert, T. J. (1995). Improving software quality through a novel testing strategy. *Proceedings of the 19th Annual International Computer Software and Applications Conference*, 224–229.

Zhang, S., Liu, X., & Deng, Y. (1997). Software quality metrics methodology and its application. *Beijing Hangkong Hangtian Daxue Xuebao/Journal of Beijing University of Aeronautics and Astronautics*, 23(1), 61–67.

Zhou, Z. (1987). Models for reliability and fail-safety evaluation of micro-computer systems. *Microelectronics and Reliability*, 27(5), 839–846.

Zhu, H., Zhang, Y., & Huo, Q. (2002). Application of hazard analysis to software quality modelling. *Proceedings of the 26th Annual International Computer Software and Applications Conference*, 139–144.

Zimmerman, D., Slater, M., & Kandall, P. (2001). Risk communication and usability case study: Implications for Web site design. *Proceedings of the IEEE International Conference on Professional Communication*, 445–452.

Zultner, R. E. (1994). Software quality function deployment: The first five years— Lessons learned. *Proceedings of the ASQC 48th Annual Quality Congress*, 783–793.

Zweben, S. H. (1992). Evaluating the quality of software quality indicators. *Proceedings of the Twenty-Second Symposium on the Interface, Computing Science and Statistics*, 266–269.

Index

A

Absorption law, 15
Accidents, computer system, 3
Advanced Research Projects Agency
    Network (ARPANET), 127
Air Force model for fault density
    prediction, 86–88
Air Force study, software costs, 183
AQAP-14, 112
Ariane space rocket, 2, 116
Asoka, 13
Associative law, 15
Availability, computer system, 3

B

Bathtub hazard rate curve, 22, 29–31
Behavior, human, 55
Boole, George, 15
Boolean algebra
    laws of, 15, 16
    origins, 15
Bottom-up software testing, 96–97
Bridge networks, 42–44
Bugs, software. *See also* Software safety
    catastrophic incidents, 115, 116
    examples, 115, 116

C

Cardano, Girolamo, 13
Cause-consequence diagrams, 132
CERN. *See* European Laboratory for
    Particle Physics (CERN)
Checklists, use during software
    development, 153
Code of Hammurabi, 51
Code walk-through, 131, 132
Cognitive walkthrough, 151–152
Commutative law, 15
Computer system life-cycle costing
    algorithmic models, 190–191
    analytic models, 191

composite models, 192
costing process, 184–185
elements of life cycle, 185–187
estimating models, 190–191
linear models, 191–192
maintenance costs, 182–183
mathematical models, 177–179, 180,
    187, 188–190
multiplicative models, 191
overview, 177
servicing labor costs, 181–182
tabular models, 192
Computer systems
    deaths resulting from failures of, 2
    failures of, famous, 2
    life-cycle; *see* Computer system
        life-cycle costing
    reliability; *see* Reliability,
        computer systems
    usability, 4
Consensus recovery block, 94
Correctedness, proof of, 132, 134
Cumulative distribution function, 17–18

D

Data structure design, 95
Debugging
    definition of, 3
Distributive law, 15
Dynamic fault tolerance, 64

E

E-commerce, 160
Electronic commerce (EC), 136
Error, human. *See* Human error
Errors, software
    causes, 118
    classification I errors, 119–120
    classification Il errors, 120, 121
    common types of, 119
    cumulative failure profile, 122–123

Milton Keynes UK
Ingram Content Group UK Ltd.
UKHW040105071024
449327UK00019B/818

9 780367 380006